D1251036

ENVIRONMENTAL DATA HANDLING

ENVIRONMENTAL SCIENCE AND TECHNOLOGY

A Wiley Interscience Series of Texts and Monographs

Edited by ROBERT L. METCALF, *University of Illinois*
JAMES N. PITTS, Jr., *University of California*
WERNER STUMM, *Eidgenössische Technische Hochschulen, Zurich*

PRINCIPLES AND PRACTICES OF INCINERATION
Richard C. Corey

AN INTRODUCTION TO EXPERIMENTAL AEROBIOLOGY
Robert L. Dimmick

AIR POLLUTION CONTROL, Part I
Werner Strauss

AIR POLLUTION CONTROL, Part II
Werner Strauss

APPLIED STREAM SANITATION
Clarence J. Velz

ENVIRONMENT, POWER, AND SOCIETY
Howard T. Odum

PHYSICOCHEMICAL PROCESSES FOR WATER QUALITY CONTROL
Walter J. Weber, Jr.

ENVIRONMENTAL ENGINEERING AND SANITATION
Joseph A. Salvato, Jr.

NUTRIENTS IN NATURAL WATERS
Herbert E. Allen and James R. Kramer, Editors

pH AND pION CONTROL IN PROCESS AND WASTE STREAMS
F. G. Shinskey

INTRODUCTION TO INSECT PEST MANAGEMENT
Robert L. Metcalf and William H. Luckmann, Editors

OUR ACOUSTIC ENVIRONMENT
Frederick A. White

ENVIRONMENTAL DATA HANDLING
George B. Heaslip

THE MEASUREMENT OF AIRBORNE PARTICLES
Richard D. Cadle

ANALYSIS OF AIR POLLUTANTS
Peter O. Warner

ENVIRONMENTAL DATA HANDLING

GEORGE B. HEASLIP

A WILEY-INTERSCIENCE PUBLICATION

JOHN WILEY & SONS

NEW YORK • LONDON • SYDNEY • TORONTO

Library of Congress Cataloging in Publication Data:

Heaslip, George B. 1942–
Environmental data handling.

(Environmental science and technology)
"A Wiley-Interscience publication."
Bibliography: p.
Includes index.
1. Remote sensing systems. 2. Automatic data collection systems. 3. Transducers. 4. Electronic data processing—Environmental engineering. I. Title.

TD153.H4 621.36′7 75-12779
ISBN 0-471-36672-2

Printed in the United States of America

10 9 8 7 6 5 4 3 2 1

TO JANET

SERIES PREFACE

Environmental Sciences and Technology

The Environmental Sciences and Technology Series of Monographs, Textbooks, and Advances is devoted to the study of the quality of the environment and to the technology of its conservation. Environmental science therefore relates to the chemical, physical, and biological changes in the environment through contamination or modification, to the physical nature and biological behavior of air, water, soil, food, and waste as they are affected by man's agricultural, industrial, and social activities, and to the application of science and technology to the control and improvement of environmental quality.

The deterioration of environmental quality, which began when man first collected into villages and utilized fire, has existed as a serious problem since the industrial revolution. In the last half of the twentieth century, under the ever-increasing impacts of exponentially increasing population and of industrializing society, environmental contamination of air, water, soil, and food has become a threat to the continued existence of many plant and animal communities of the ecosystem and may ultimately threaten the very survival of the human race.

It seems clear that if we are to preserve for future generations some semblance of the biological order of the world of the past and hope to improve on the deteriorating standards of urban public health, environmental science and technology must quickly come to play a dominant role in designing our social and industrial structure for tomorrow. Scientifically rigorous criteria of environmental quality must be developed. Based in part on these criteria, realistic standards must be established and our technological progress must be tailored to meet them. It is obvious that civilization will continue to require increasing amounts of fuel, transportation, industrial chemicals, fertilizers, pesticides, and countless other products and that it will continue to produce waste products of all descriptions. What is urgently needed is a total systems approach to modern civilization through which the pooled talents of scientists and engineers, in cooperation with social scientists and the medical profession, can be focused on the development of order and equilibrium to the presently disparate segments of the human environment. Most of the skills and tools that are needed are already in existence. Surely a technology that has created such manifold environmental problems is also capable of solving them. It is our hope that this Series in Environmental Sciences and Technology will not only serve to make this challenge more explicit to the established professional but that it also will help to stimulate the student toward the career opportunities in this vital area.

Robert L. Metcalf
James N. Pitts, Jr.
Werner Stumm

PREFACE

The subject matter contained in the chapters of this book was designed to serve as a text for the fundamentals of remote sensor data acquisition and data-handling courses and to act as a reference guide for readers who require a basic knowledge of transducer data and the manner in which analog and digital systems perform data processing and final data display.

Each chapter is written expressly for personnel and students who do not necessarily possess a strong test or remote sensor data-handling background but instead need to become familiar with raw test data formats and modern techniques especially tailored for use by the environmental or test data analyst.

For the reader concerned primarily with aircraft and/or satellite remote sensing, chapters on analog-and-digital data acquisition should prove to be of high value, for many well-executed remote sensor programs require ground truth data (acquired by transducer) for comparison with plane or satellite data to establish the validity of remote sensor findings. Also, because aerial or satellite data are available only periodically (e.g., every 18 days from the NASA Earth Resources Technology Satellite), some transducer data are often necessary when the required frequency of data acquisition is high.

For the reader whose chief interest lies in transducers and analog or digital data acquisition and processing systems the chapters on remote sensing discuss data-base design considerations and a variety of graphics data display techniques, many of which are potentially applicable to his work.

Because numerous texts on the subject of digital computers are available today, this topic is not discussed in depth. Instead, a large part of this book is devoted to basic transducer and sensor data-recording methods and to descriptions of output data useful to those concerned with subjecting physical items and man-made structures to test or to those involved in monitoring and assessing natural and cultural features with ground, aircraft or spacecraft sensors.

Particular emphasis is placed on vibration analysis, acoustic data, and the manner in which remote sensor data are collected and machine processed for studies in agriculture, marine resources, hydrology, land use, geology, and the general environment. Analog and digital graphics systems are highlighted and one chapter is devoted to microfilm, the ideal and cost-effective storage medium for large quantities of data printed normally on paper or as images on photographic paper.

The reader who masters the fundamentals of each of the chapters pertinent to his work will acquire knowledge of those factors that are a prerequisite to the effective use of spaceborne, airborne, and ground-acquired survey data.

GEORGE B. HEASLIP

East Islip, New York
March 1975

CONTENTS

ENVIRONMENTAL DATA HANDLING

1

INTRODUCTION TO TRANSDUCERS AND REMOTE SENSORS

The collection of data that relate to the physical, chemical, optical, or biological characteristics of both natural features and man-made bodies may be achieved in three ways:

1. *In situ electronic transducers.* Devices attached to or placed in a body being monitored. Outputs are electrical signals with voltage variations analogous to the physical variations measured and suitable for routing to meters, oscilloscopes, chart recorders, and analog or digital magnetic data tape recording systems.

Literally hundreds of transducer types are available, each of which is designed for a unique application; for example, strain gages can be mounted in an aircraft during a flight test, accelerometers monitor rocket motion, seismic transducers monitor earth vibrations, the hydrophone can detect undersea acoustics, and current meters convert water movement characteristics to an electric signal suitable for recording. These are just a few.

2. *Field or laboratory observations.* Optical, chemical, physical, biological, or photographic observations of body characteristics made at the test site or at the laboratory to which collected samples are transferred. Results are usually manually generated logs or listings.

3. *Remote sensing.* The detection of body characteristics by sensors

connected in no way to the body or bodies under observation. A good example is the aircraft or satellite multispectral scanner which detects spectral reflectance and emittance from the surface areas of the earth.

In situ transducers play a major role in the testing of manufactured components and structures (e.g., aircraft and spacecraft) and in data acquisition systems designed to record the characteristics of land, water, and living complexes. Output data may be routed directly to analysis and display devices, recorded on tape for later analysis, or routed by FM telemetry to distant data processing sites.

Telemetry not only assists in relaying *in situ* data from plane or spacecraft back to earth (and vice versa) but also in transferring earth-resources remote sensor data from satellite to ground. Earth-resources satellites can record and transmit spectral data and are capable of receiving and relaying transducer data acquired and transmitted to satellite by ground-based data-collected platforms.

The manner in which physical data are actually collected, conditioned, and recorded is described in the first chapters of this book. Later chapters discuss FM and PCM data acquisition and processing systems. Analog data display devices are covered, as are the analog-to-digital and digital-to-analog processes. The uses of microfilm, as it relates to test and earth-resources data processing, are also highlighted.

Heavy emphasis is placed on fundamentals of dynamic analysis and earth-resources data handling. A knowledge of these topics is a prerequisite for the reader who is concerned with the analysis of acoustic and vibration data and with the application of transducer and remote sensor technology to studies of natural, cultural, and environmental resources. A glossary of key environmental and test data-handling terms has also been prepared.

The primary purpose of an environmental or test data-handling plan is to ensure the extraction and proper formatting of meaningful data. Too often too many data undergo extensive and costly electronic processing and the extremely valuable graphics display techniques are not used. Included, therefore, are discussions of the various systems and test data processing options.

2

TRANSDUCERS AND REMOTE SENSORS

When speaking of environmental conditions that must be measured and recorded, we are usually referring to shock, vibration, temperature flow, or other parameters that can be monitored by the *in situ* transducer. If we are interested in the status of land or sea, we are sometimes concerned with the nature of optical emissions and reflected energy from earth features which can be measured from plane or satellite by remote sensors—that is, multiband cameras and/or the multispectral scanner.

Both the transducer and remote sensor are active in an overall acquisition system; each converts measured energy (i.e., physical or optical) into an electrical signal.

The total data acquisition system contains several parts:

1. *The sensor* detects phenomena occurring on land and sea, in the air, or on board a ground vehicle, aircraft, or spacecraft.

2. *The signal conditioner* amplifies or otherwise conditions sensor outputs for suitable presentation to an analog or digital recording system.

3a. *The voltage-controlled oscillator* converts sensor voltage to a frequency-modulated (FM) signal suitable for tape recording or transmission by telemetry to a distant data processing site.

3b. *The analog-to-digital converter* repetitively samples sensor output data and transforms each sample into a digital waveform suitable for computer entry or tape recording.

4. *The magnetic data tape recorder* records analog or digital data at the test site, in a plane, or at a telemetry data receiving site.

The sensor (i.e., a transducer or remote sensor) is the first component in a total system. Its data are tape recorded and the tapes then serve as input data to an analog or digital data processing system.

For the most part transducer and sensor outputs are analog or, more specifically, continuous electrical signals that vary proportionately with the physical phenomena being measured.

When vast quantities of data are obtained, it is often necessary to tailor them for entry into a digital computer. When only a few items (i.e., parameters) are to be processed, analog techniques alone usually suffice. Analog systems, both data acquisition and data processing, deal with *continuous* data. Digital systems, on the other hand, require action on analog *samples* which undergo pulse-code modulation (PCM) as a prerequisite to digital computer processing. When a measured parameter, such as an underwater acoustic signal or a vibratory signal from an airplane, is high in frequency, conversion from an analog to a digital format becomes difficult and/or expensive. For this reason *hybrid systems* which accept data from sensors and route low-frequency parameter data to analog-to-digital converters and high-frequency parameters to the analog voltage-controlled oscillator, are sometimes used. The data that result are recorded onto a hybrid tape whose contents (analog and digital) can be routed later to separate systems at the data processing site.

It is important to note that data processing techniques are essentially the same for transducer and remote sensor data. Often sensor data from aircraft flight tests undergo the same kind of processing as undersea or earth-resources data. Thus it must be remembered that many of the discussions presented in this book apply simultaneously to several study fields.

In some cases (e.g., in the water-resources survey area) it is necessary to obtain and process electronically not only aircraft and satellite data but also *in situ* data that pertain to water temperature, turbidity, salinity, dissolved O_2, and so on. The balance of this chapter is devoted to a discussion of several *in situ* transducers and highlights two major remote sensing systems—the thermal infrared and multispectral scanners.

POTENTIOMETRIC TRANSDUCER

The potentiometric transducer operates by Ohm's law, by which physical energy is transferred to a slider in a potentiometer circuit device. Variations in physical energy (e.g., pressure) change the ratio of output resistance to total circuit resistance. According to Ohm's law, output voltage varies proportionately with resistance changes. Also, in the pressure transducer (Figure 1) output voltage is directly proportional

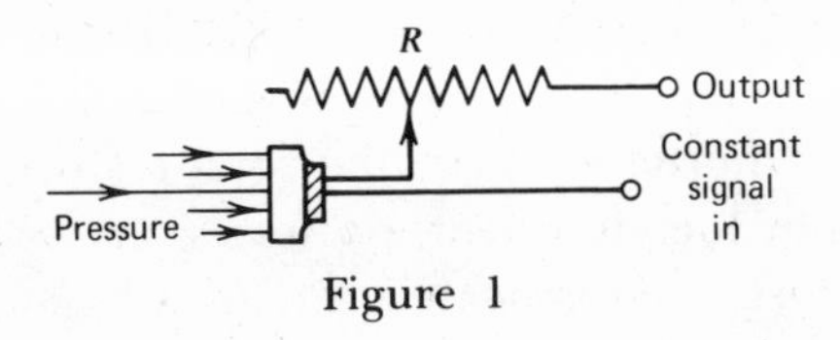

Figure 1

to pressure applied to the sliding tap mechanism. Circuit resistance decreases as pressure moves the sliding tap (arrow) to the right. In the potentiometer pressure transducer pressure variations generate an analogous change in resistance, hence in output voltage.

Additional examples follow. The first device illustrated (Figure 2) can measure the effects of air movement effectively and the second, an accelerometer, yields force and acceleration results.

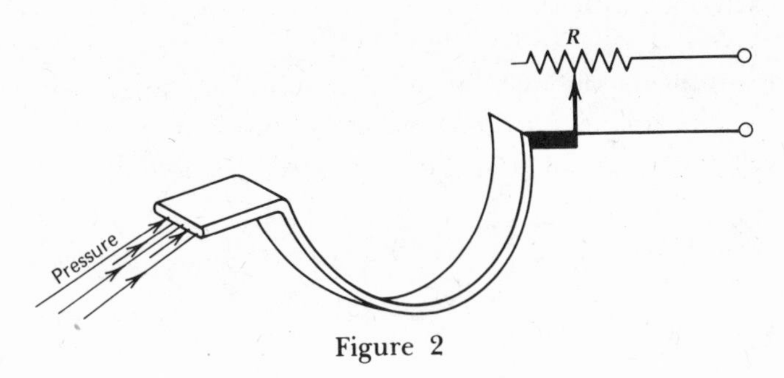

Figure 2

The Accelerometer

With the accelerometer (Figure 3), acceleration or vibration causes movement of the sliding mass. As mass displacement X is completed, proportional changes occur at the transducer's output terminal. Also, according to the laws of physics, force = (mass) × (acceleration) = $-KX$, where K is the spring constant. From this equation we may

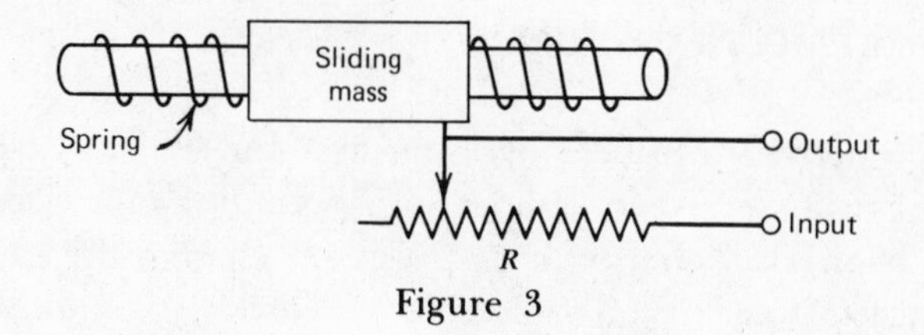

Figure 3

derive acceleration = $-KX$/mass, and because X is directly proportional to output voltage we can then ascertain acceleration values.

Another "resistance type" of transducer is that in which the output signal depends on changes in the resistance of an internal element. Examples are the thermistor and the strain gage. The thermistor, used in temperature measurement, possesses a resistor element whose ohms rating decreases as it is heated. The strain gage operates on the principle that its resistance to an electrical signal also changes as wire length is changed (by strain).

In summary, the operational principle for resistive transducers states that when a constant electrical signal is applied and when mechanical changes cause resistive changes within it output voltage is directly proportional to those mechanical changes.

MAGNETIC TRANSDUCER

These transducers operate on another principle of physics; that is, whenever a permanent magnet passes a coil, it induces an electrical signal. A basic vibration transducer (Figure 4) contains a magnet sus-

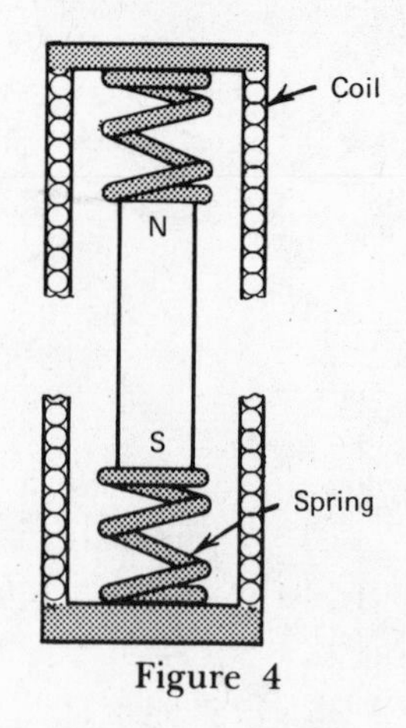

Figure 4

pended in a coil of wire. When vibrations cause movements of the magnet, voltage changes occur in the coil. More precisely stated, vibratory changes cause magnetic flux changes. Liquid flow rate may be determined with a magnetic transducer called the flow meter. In this device a magnet placed on a turbine blade in the flow path will induce voltage changes in a nearby coil as changes in blade revolution rate occur. Voltage fluctuations in the magnetic flow meter are proportional to liquid or air flow changes in the tube containing the turbine blade (Figure 5).

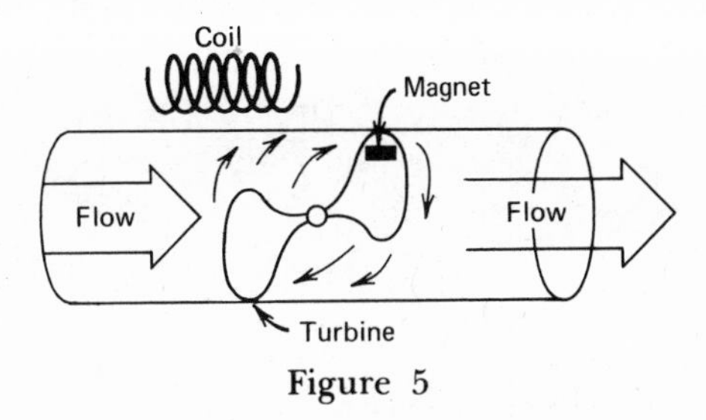

Figure 5

PIEZOELECTRIC TRANSDUCER

Transduction is achieved in this device by the unique piezoelectric characteristics of certain crystals and salts. Usually quite lightweight in nature, the piezoelectric sensor (Figure 6) produces a charge across its

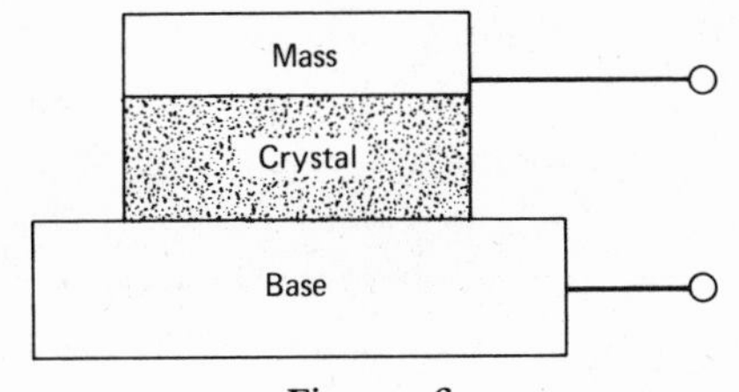

Figure 6

plates, the mass and base, that is proportional to the force applied to the crystal. Output voltage relates to charge generated and crystal capacitance according to this formula: volts = generated charge/ transducer capacity = Q/C.

In addition to the *in situ* transducers already discussed are several others. the *electrochemical transducer* makes use of a chemical change to monitor the input parameter and also the fact that output electrical signals are identically proportional to chemical and physical parameter variations. Another type, the *inductive transducer,* is one in which physical changes are conveyed by changing the device's actual inductance rating. Still others include the *variable capacitance,* the *photoelectric,* the *force balance,* and the *ionization.*

No matter what the type, each converts energy from one form to another while retaining the amplitude characteristics of the energy being converted. Remote sensors, discussed in the chapters that follow, perform essentially the same energy-to-voltage conversion.

The transducer and remote sensor act as first elements in data acquisition systems. Chapter 4 discusses conversion of signal-conditioned sensor outputs into frequency-modulated (FM) carriers. Later, in Chapter 9, we describe the conversion of analog sensor data to a digital or computer compatible format.

THE INFRARED SCANNER (Thermal Mapper)

In infrared sensing heat radiation from earth surfaces is detected and recorded. It must be noted that all earth objects with temperatures greater than -273°C emit thermal energy and emittance relates to the characteristics of each object's surface.

Recording of infrared data is usually acheived with an infrared line scanner, a device capable of recording information in the spectral region above 1.0 μ. Multiband cameras can record short wavelength data but are ineffective at about 1.0 μ and beyond.

The actual scanner functions like a television receiver in that it produces image data in a series of scan lines. Usually mounted in a spacecraft or aircraft, its temperature-sensitive components scan the terrain (by a rotating mirror) in strips perpendicular to the platform line of flight. Mirror images are routed to a detector, amplified, and then presented to a film and/or tape recording system (Figure 7). Emissions are measured from the red end of the visible spectrum where reflective energy dominates (approximately 0.7 μm) to the thermal or emissive portion (3 to 15 μm) (Figure 8). In the reflective region (0.7 to 3 μm) sun radiation is more dominant than self-emissions. In the thermal IR zone (3 to 15 μm) the reverse is true. During the day when sunlight is prevalent the 0.7 to 3 μm band

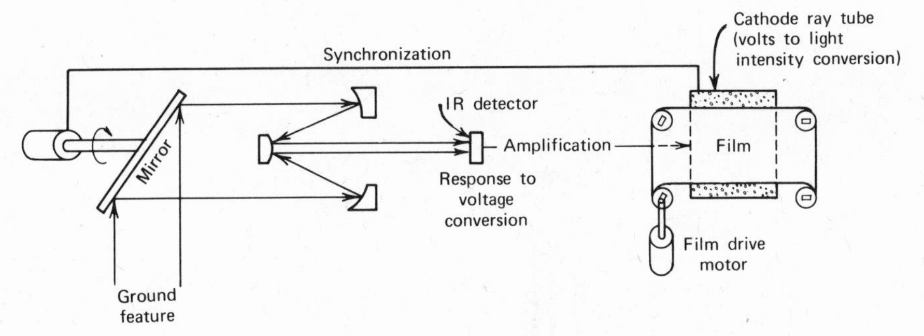

Figure 7. The IR scanner/recorder system.

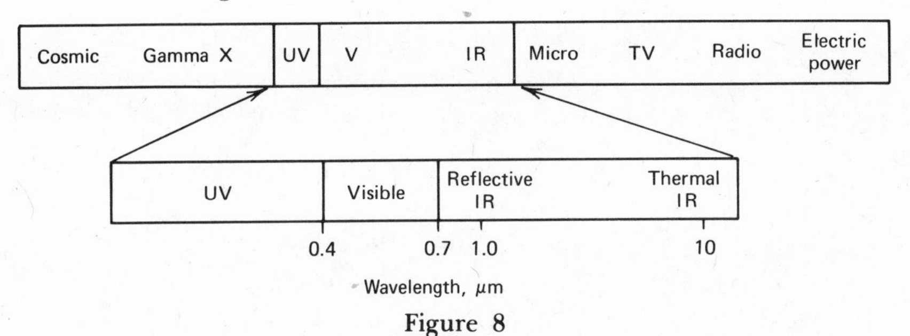

Figure 8

offers little earth-feature data of use to the interpreter; at night, however, energy detected in this band relates strongly to earth-feature emission and not to scattered sun radiation.

Selection of an adequate IR band must be based primarily on expected cloud and haze cover and atmospheric transmission characteristics because some attenuation occurs. Atmospheric H_2O and CO_2 reduce the amount of energy that the aircraft or spacecraft IR sensor should record. For many applications two "windows," about 2 to 5μm and 8 to 15 μm offer good ground-to-sensor transmission (Figure 9). To prevent "underlap" of scan lines in imagery, aircraft velocity, and altitude must be carefully selected. Selection is based on the equation

$$r \geqslant V/HlnB$$

Where r = scanner rotational rate, l = number of lines swept by mirror face, n = number of mirrors, B = angular size of resolution element, V = velocity, H = altitude.

Example. For one of the commercially available sensors (i.e., the Daedalus scanner) the line scan rate is 120/sec and the field of view is 2.5×10^{-3} rad *or* 1.7×10^{-3} rad (i.e., selectable field of view).

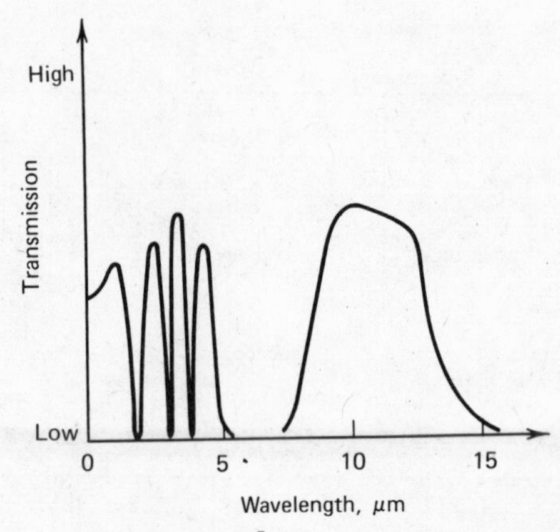

Figure 9. Transmission loss in "nonwindow" areas is due chiefly to water vapor and CO_2, each of which possesses strong infrared absorption bands.

By using $\ln Br = V/H$ = line scan rate × instantaneous field of view, we obtain (for the two fields of view):

$$120 \times 2.5 \times 10^{-3} = 0.204 = V/H \text{ (detector 1)}$$

$$120 \times 1.7 \times 10^{-3} = 0.3 = V/H \text{ (detector 2)}$$

From these relationships V and H may be selected (Figure 10).

Earth features, including water bodies, land, and plants, radiate IR energy as a function of their temperatures. Temperature, in turn is related to a feature's heat capacity, thermal conductivity, and moisture content and to environmental factors such as sky cover, wind, precipitation, and land topography.

Analysis of IR data, whether by analog imagery or computer-generated gray scale or contour maps is by no means a straightforward operation. The interpreter must understand several factors:

1. Sensor-error characteristics.
2. Aircraft motion error (from a straight line parallel to the earth).
3. Nonuniform signals from ground features which are in fact at identical temperatures.

. . . This phenomenon develops because the ability of objects to radiate is not uniform. Two features at identical temperatures could appear in different gray tones on an IR image.

4. Night versus day temperature changes are often sizable for some features but small for others (e.g., high for asphalt roads, low for water).

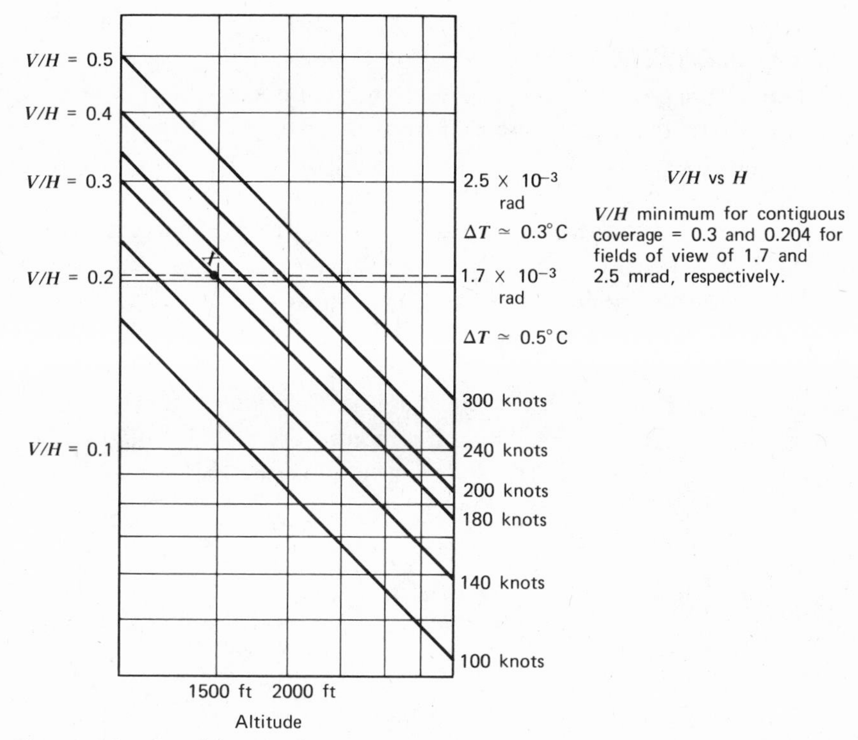

Figure 10. Daedelus IR scanner. Example: Assume that aircraft speed is 180 knots and that a 1.7-mrad field of view is desired. From the graph (reference point X) the minimum aircraft altitude must be approximately 1490 ft.

Sensor spatial resolution is a function of the ratio of IR detector size to optics focal length, d/f; d is usually fixed to about 0.1 mm because of the mechanics of construction and f, if chosen too large, causes a smaller field of view with degraded temperature sensitivity. High-quality sensors are generally equipped with a "blackbody" reference whose known temperature serves as a calibration signal to which earth-feature data can be compared.

In order to produce IR imagery (paper or film) or to effect computer contouring or gray scale shading, scanner synchronization signals must be recorded along with detector-measured radiation. Synchronization data, together with known aircraft speed, enables data processing personnel to adjust the output recording system so that a proper number of scan lines per inch will appear on final imagery. Density, on images, directly relates to effective radiation temperature;

positive prints show warm areas as light shades of gray and cool areas as dark shades. The reverse is true on negative print images.

Infrared images can be generated by an onboard film system or later by processing taped data and presenting the results to a cathode ray tube (for filming) or fiber optics oscillograph recorder. Although IR images resemble photographs, they can contain errors, several of which are based on the nature of scanning from a moving system (e.g., aircraft) which is subject to roll, pitch, yaw, and wind.

Corrections for motion error such as roll can be made if roll factors are recorded on tape with data (for later processing) or if the onboard system has roll compensation circuitry. Also, unless corrected for, another type of error can develop because image scale along the line of flight generally differs from that across the path of flight. The cause of scale variations across a scene is illustrated in Figure 11.

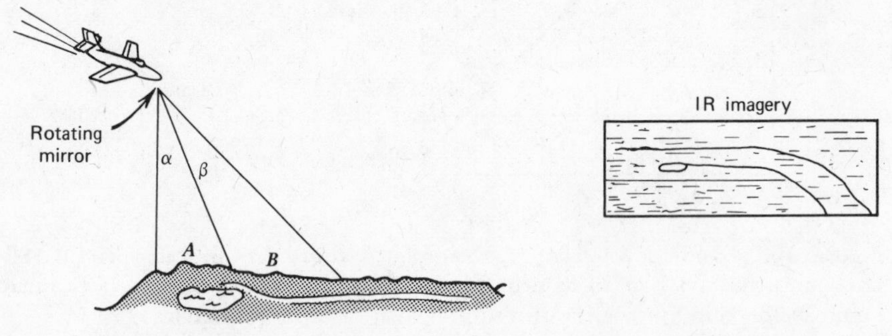

Figure 11. Although $\alpha = \beta$, $B > A$ or although the mirror rotation is a linear function of angle, image roll-off occurs because terrain coverage (in size) is not.

When detector data are routed to a tape recorder, analog or digital computer programs can be used to correct data for scale-variation error and render them suitable for both temperature and distance measuring. Tape data can also be presented to a color cathode ray tube for color enhancement.

Whether of the image (or "strip map") format or computer enhanced display, thermal IR data have been applied in numerous areas which include but are certainly not limited to the following:

1. Volcano surveillance.
2. Soil-moisture survey.
3. Fresh-water outfalls detection.

4. Crop species and crop-stress survey.
5. Thermal-plumes survey at power plants and industrial sites.
6. Irrigation survey.

Although mounted primarily in aircraft, it is anticipated that thermal scanners will be installed in future earth-resources survey satellites. Figures 12 and 13 are actual samples of aircraft IR imagery of

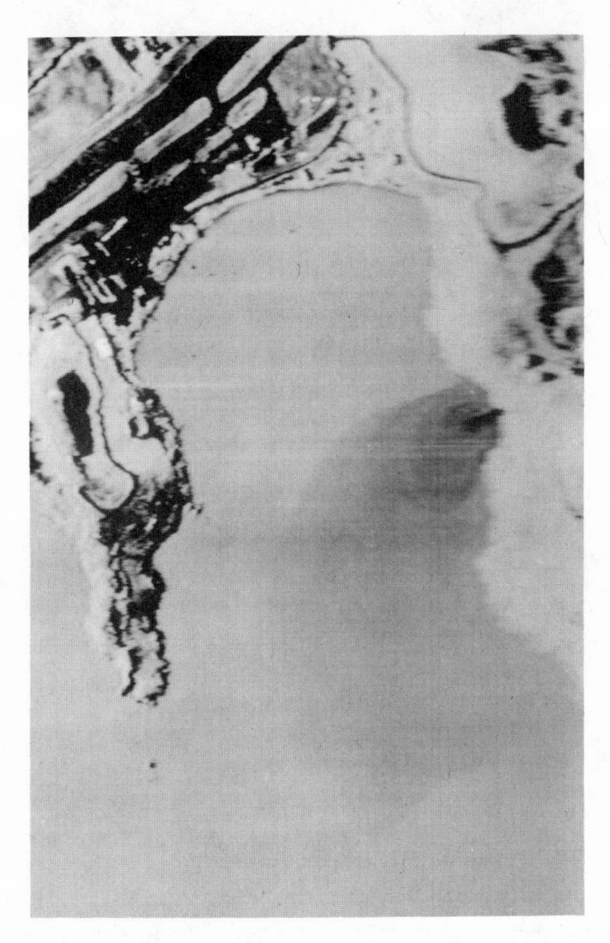

Figure 12

Lindbergh Bay, St. Thomas, where a large thermal plume appears. Here imagery was generated from an analog tape and presented to a fiber optics oscillograph recorder.

Figure 13

THE MULTISPECTRAL SCANNER

The aircraft thermal infrared scanner, discussed earlier, detects radiant flux in a thermal IR spectral band, typically in the 8 to 14 μ atmospheric window. The data that result, when displayed in an imagery or computer map format, can be highly useful; for example, in the monitoring and assessment of thermal plumes and in the location of fresh-water outfalls.

In cases in which water quality, vegetation status, geological features, and land-use patterns are to be studied from aircraft, satellite, or space station heights use of the multispectral scanner is recommended. The MSS differs from the IR scanner in that data from more than one spectral passband are recorded.

Depending on the type of monitoring to be performed, different passbands become useful. In water-quality studies a blue-green band permits the scanning of water characteristics such as sediment plumes. The near IR band, on the other hand, clearly indicates land/water boundaries. For certain agricultural applications yellow band data and thermal IR data are needed. Most of today's well-designed remote sensing programs, as they relate to earth-resources monitoring and mapping, require multiband data.

The large amount of data collected by the multispectral scanner, whether in aircraft or spacecraft, usually requires that data be converted to a digital format and processed by a digital computer. (Basics of the analog-to-digital process are discussed in Chapter 9.) Computer software (i.e., programs) which operate on such data is often structured by the programmer to produce data in a graphics or map format tailored specifically for the analyst and/or decision maker.

Figure 14 is a block diagram of the basic aircraft multispectral scanner. Satellite scanners are similar but can relay detected results via telemetry to a ground receiving site.

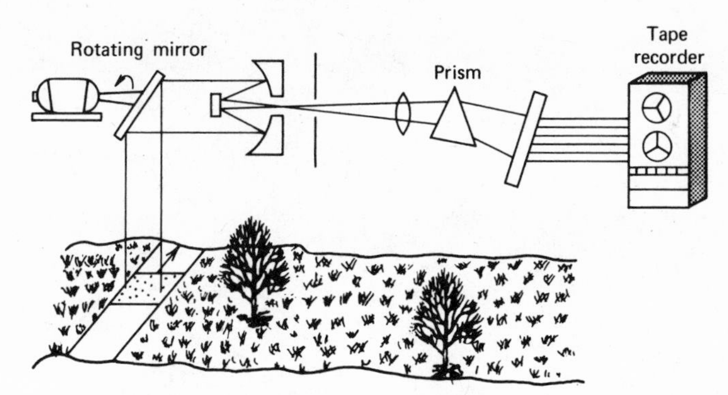

Figure 14. The multispectral scanner.

Of the two types of remote sensor, the *passive* and the *illuminate,* the multispectral scanner, like the IR scanner, is classified as passive. More specifically, the passive sensor records emissions and reflections from below. The illuminate sensor system emits an electromagnetic signal and then records the energy reflected back; radar is an example.

When working with MSS data, one must remember that inputs to the sensor, and the sensor detector outputs that result, are dependent on a variety of conditions:

- Sun angle.
- Reflectance properties and spatial distribution of objects in the sensor field of view.
- Atmospheric transmission characteristics.
- Sensor location with respect to ground features.

The total illumination on ground objects constitutes not only solar illumination but also radiation from other parts of the sky and reflections from objects in proximity to the feature(s) under study. Pollution, haze, and clouds are other variables that affect final sensor outputs.

In the agricultural or natural vegetation survey variation in reflectance generally depends on the state of each type of vegetation, but soil condition, sun angle, and topographic conditions must also be considered.

Although attenuation by absorption is not a major problem for the MSS operating in the visible region (i.e., 0.4 to 0.7 μm), scattering can be a cause. The key message here is to bear in mind always that energy recorded by the MSS for a specific ground feature may or may not contain error, depending on conditions that exist at the time of the satellite or aircraft overpass.

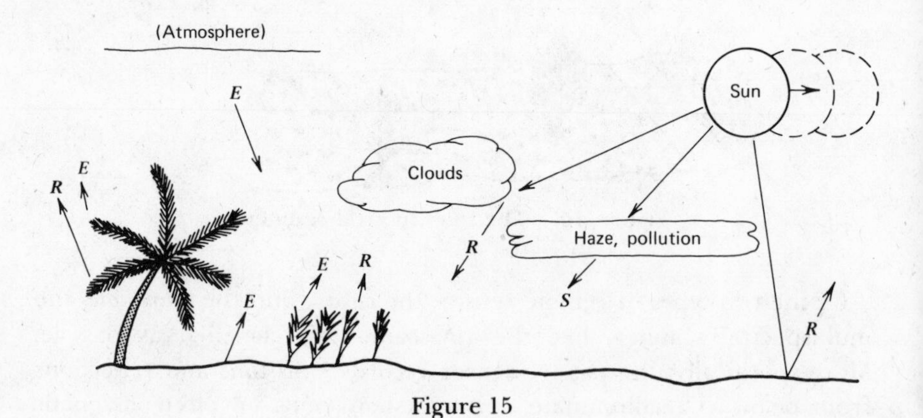

Figure 15

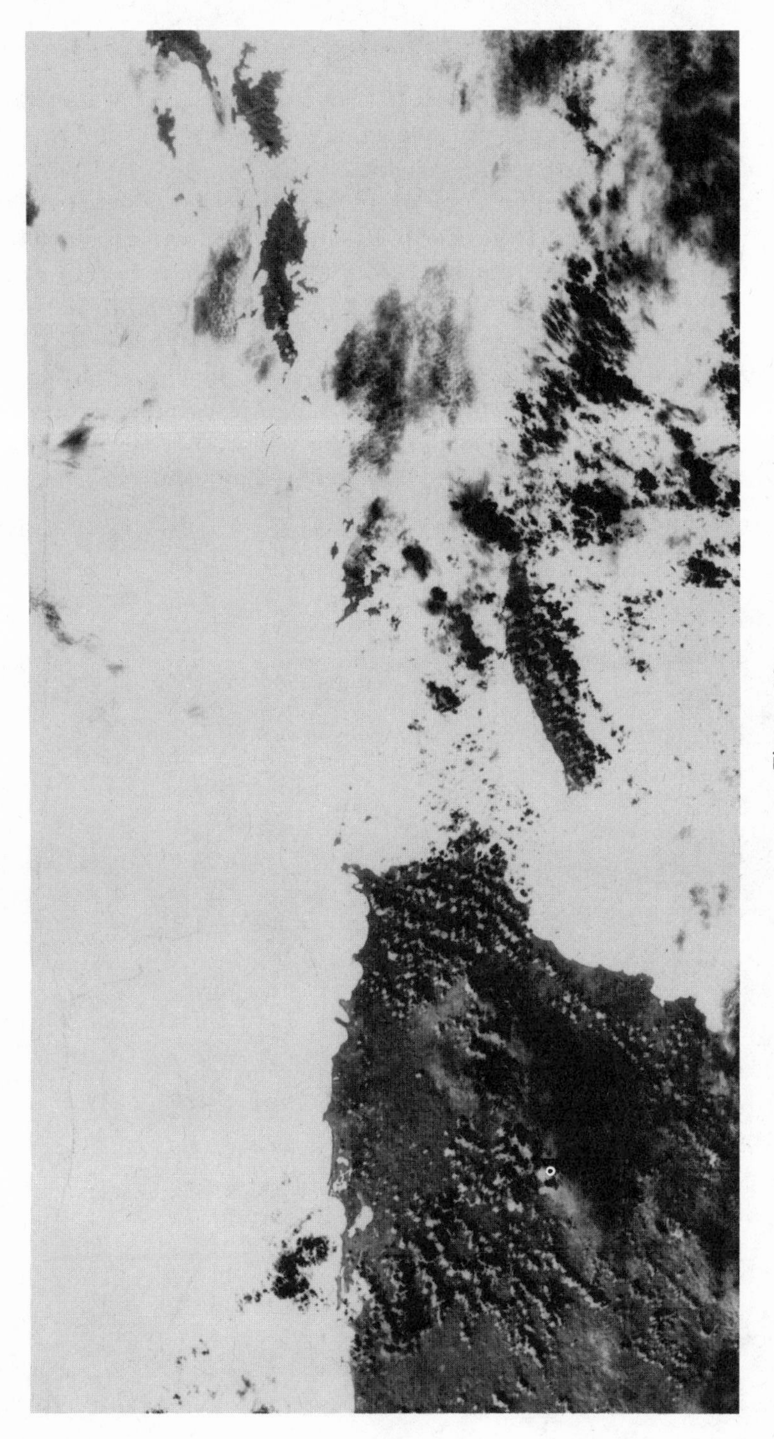

Figure 16

Figure 15 illustrates reflected (R), emitted (E), and scattered (S) electromagnetic energy. Also shown are examples of multispectral scanner data obtained from a satellite.

Figure 16 is a portion of a NASA-produced 9 × 9 in. black and white image obtained from Earth Resources Technology Satellite 1. Appearing in this scene, obtained from MSS band 6, is eastern Puerto Rico at the left and St. Thomas, St. John, and Tortola in the upper right. The dark region, to the right of center in Puerto Rico, is a cloud formation over the famous rain forest. Smaller cloud formations, seen scattered over the land mass, are black and white (cloud and shadow). One such formation has formed over western St. John.

Of the four ERTS* bands, MSS 4 is frequently the most useful in

*ERTS = Earth Resources Technology Satellite; renamed "LANDSAT" in early 1975.

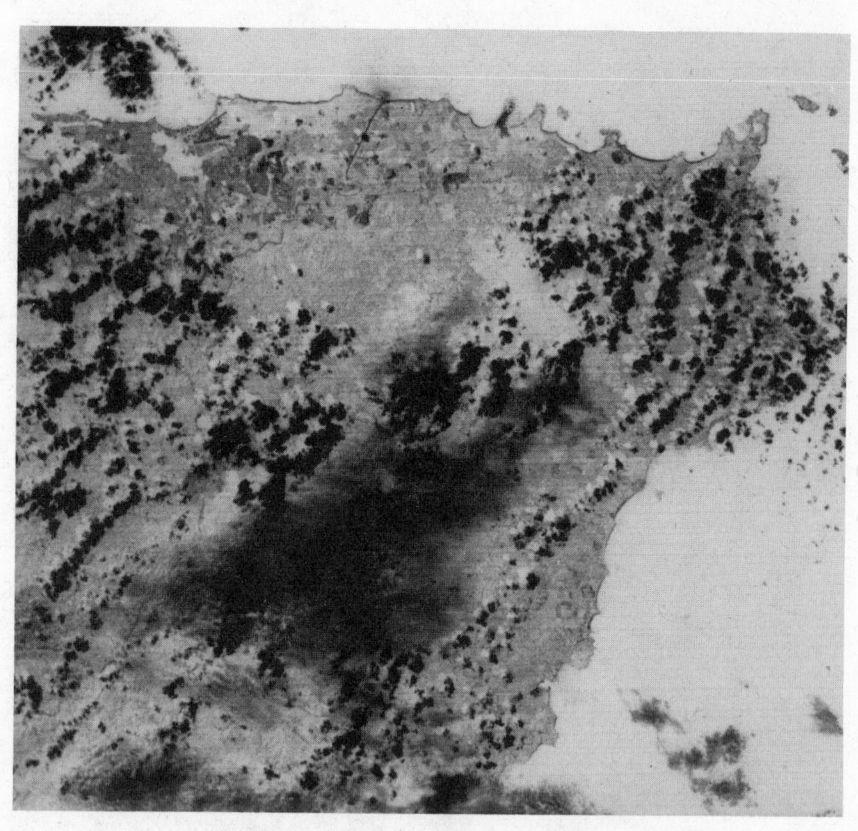

Figure 17. ERTS 1 MSS band 4. Eastern Puerto Rico.

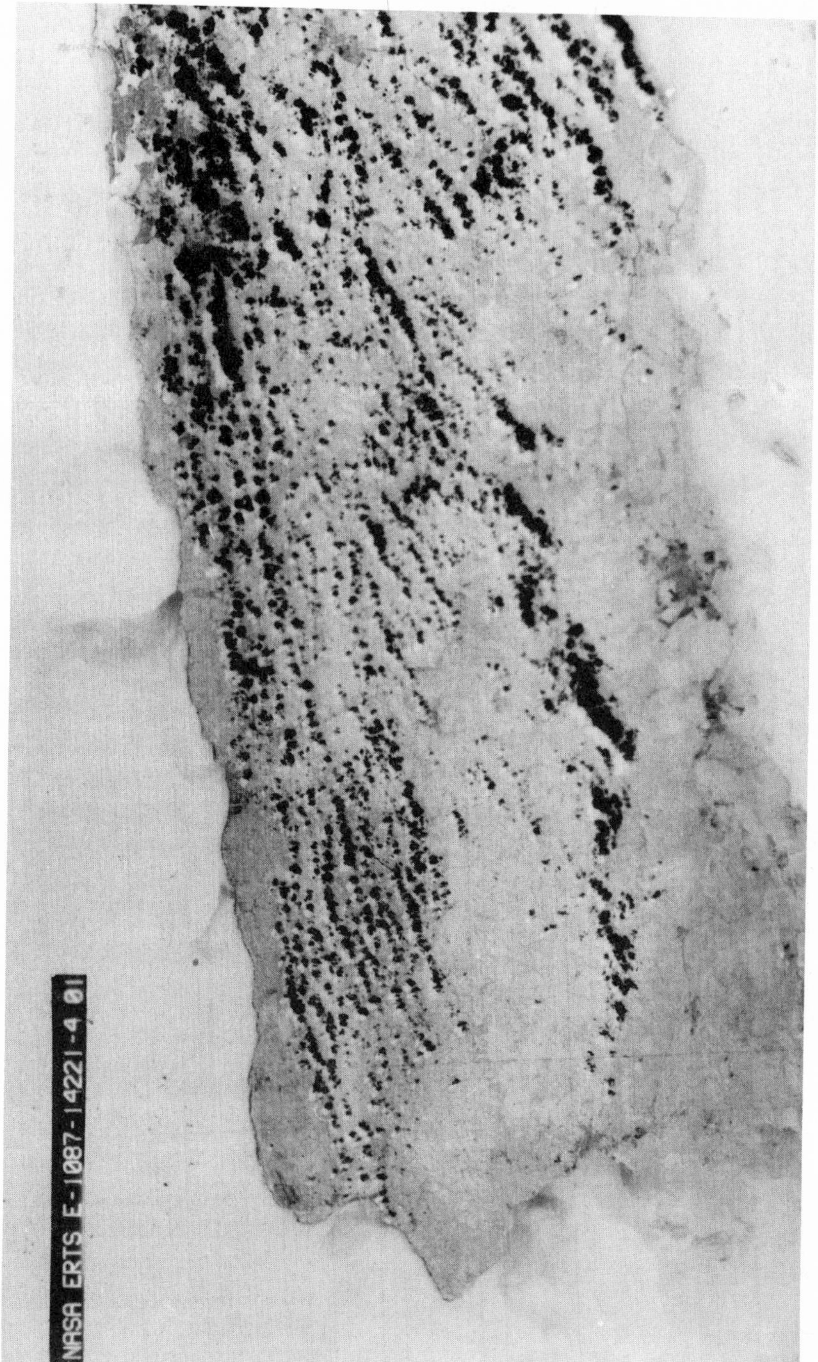

Figure 18

studies related to the coastal environment. An enlarged MSS band-4 image for the Puerto Rico area is also shown (Figure 17).

Figure 18 is a satellite multispectral scanner image in the 0.4 to 0.5 μ range. The land area shown is Puerto Rico. Note that airport runways are visible to the east of San Juan (at the upper right of the image) and that plumes in coastal waters are observable in many areas.

3

OVERVIEW OF RESULTS AND CAPABILITIES OF SATELLITE AND AIRCRAFT SENSORS

In July of 1972 NASA launched Earth Resources Technology Satellite #1 (ERTS-1). Although only the first in a series, its multispectral scanner data has already found widespread use throughout the world by those concerned with monitoring and assessing the state of natural, cultural, and environmental resources.

Multispectral scanner data were made available in several formats: as images from sensor bands 0.5 to 0.6, 0.6 to 0.7, 0.7 to 0.8, and 0.8 to 1.1 μm, as a color composite of the first, second, and fourth bands, or in digital tape format (i.e., computer compatible tapes). Band data, acquired from about 500 miles high, possessed resolution of about 80 m and each satellite scene covered an area of 185 × 185 km. Repetitive ground-area survey was possible every 18 days, cloud cover permitting. Four computer tapes recorded each scene.

Review of progress reports made by many ERTS data investigators reveals that orbital-height remote sensor data have direct application in all of the following areas:

1. Water and marine resources.
2. Environmental survey.

3. Geology.
4. Land-use monitoring and assessment.
5. Agriculture, forestry, and range resources.

ERTS, however, lacked a thermal IR channel. When such data are required, aircraft must be used. Also, when 80-m resolution was insufficient in the visible bands, aircraft multiband photography was obtained. Still ERTS images (paper or film) and the digital tape results have in several areas produced large cost savings by "telling" the aircraft where not to go. Instead of obtaining many line miles of aerial photography, satellite data have frequently shown anomalies, such as land-change patterns, and, most importantly, have pointed out specific zones in which more detail, via aircraft, was mandatory.

In general, analysis of a satellite scene can be accomplished more precisely by using the digital computer printouts (made from computer compatible tapes) than by photo-interpreting images; for example, one ERTS-1 computer printout cell or pixel (picture element) represents an area of about 57 × 79 m with 64 gray-level radiance values ranging from value 0 to saturation (black to white), whereas in the imagery enlarged to scale 1:250,000 one resolution cell has dimensions 0.3 mm and a range of 10 to 15 gray tones. Surely it is obvious that more and more emphasis will be placed on digital products that, in effect, enable an "enlargement" of the available photo products. Perhaps one of the most frustrating problems to users of satellite and aircraft sensor data is cloud cover. In the image in Figure 1 Venezuela's north coast is relatively cloud-free, but inland areas, including Caracas, are partly cloud-covered. This image, which is part of NASA Scene 1086-14185-7, contains data recorded in the near IR spectral region. Land/water boundaries show clearly and white zones indicate clouds. Cloud shadows (black) are also visible. The large water body to the left of image center is Venezuela's Lake Valencia.

Satellite sensor data have proved extremely useful in surface-water mapping but are generally poor for surveying the manner in which water is used by man. Mapping and monitoring of large rivers and lakes is possible but mapping of irrigation is not. Wetlands survey is an ERTS data application but mapping of soil moisture content cannot be readily achieved. However, the proper combination of satellite data with aircraft thermal IR line-scanner data and multiband photography can yield outstanding results.

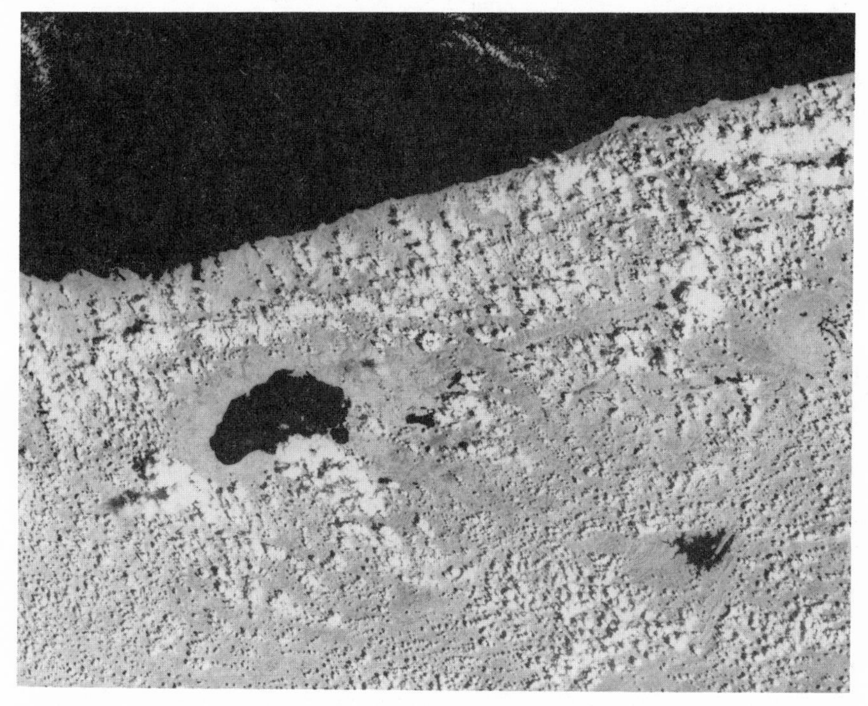

Figure 1

Aircraft thermal IR imagery is especially valuable in detecting coastal fresh-water outfalls entering major bodies of water. Fresh-water temperature usually differs from that of the open sea into which it flows. When that difference is greater than about ¼ to ½°C, the aircraft IR line scanner can be the ideal mapping tool. Near-IR data (from satellite) and/or thermal IR images (from aircraft) enable the data analyst to map land/water boundaries. Once established, spectral data from the blue-green region, as recorded by satellite, can serve as overlay data from which surface-water features and, in water of high clarity, bottom features can be observed.

Satellite data are generally useful in studying earth areas greater than 5 to 10 acres in size. When finer resolution is required, the aircraft sensor must be used. Infrared scanners aboard a low-altitude plane, for example, can achieve spatial resolution in the order of a few feet, but such data are expensive to acquire. ERTS images (185 × 185 km) are available for about 2 dollars in scale 1:1,000,000 format. Digital tapes that cover 25 × 100 nautical miles are available for about 50 dollars each.

Single-channel aircraft thermal IR line-scanner data are often FM-recorded onto a wideband carrier. When this is done, analog ¼ or ½ in. magnetic data tapes can be processed for FM data-to-image conversion. Also, analog data (i.e., demodulated FM carriers) may be analog-to-digital converted; the data that result are presented to a digital computer for gray-scale shading or contouring by temperature increments. Multispectral scanner data, on the other hand, are usually best handled by the digital computer because of the enormous quantity of information acquired by the scanner detectors.

Examples of both aircraft multiband photography (Figure 2) and

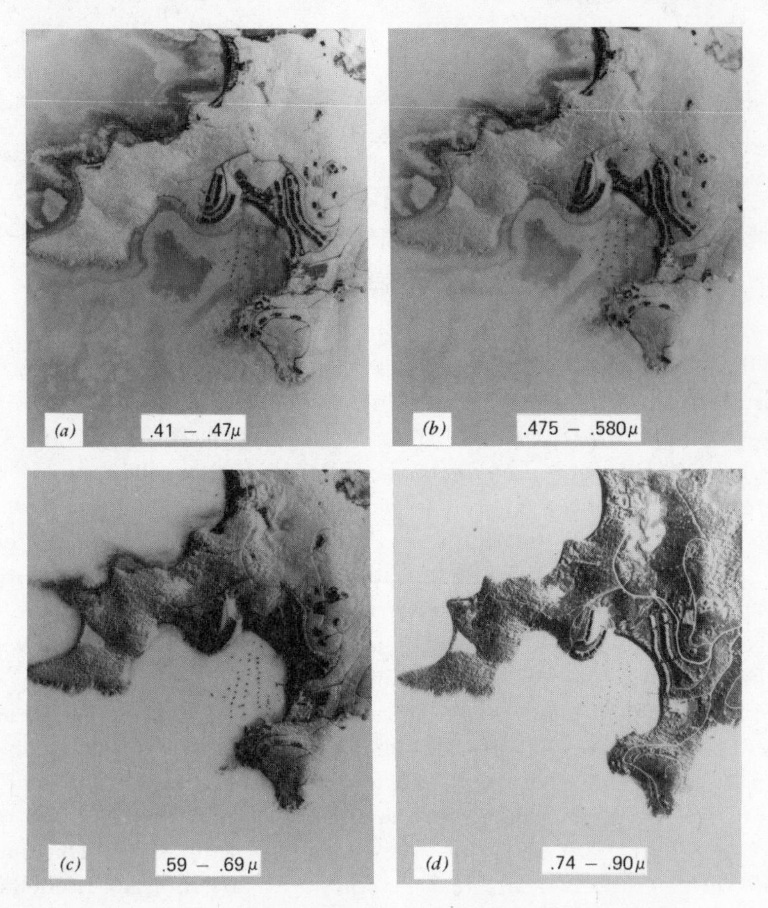

Figure 2. Aircraft multiband photography.

computer-processed satellite MSS data (Figure 3) follow. Digital MSS data may be presented in a variety of output formats, including contour and perspective plots and line-print maps. Gray-scale shading and digital data-to-image conversion is also possible. The multiband camera data shown was obtained at St. Thomas from an altitude of 6000 ft; the multispectral scanner data shown was acquired from the ERTS-1 spacecraft.

In the fields of agriculture and forestry a great deal has been done with aircraft and satellite sensors for classifying and mapping crops and ground vegetation stress. Plant leaves possess optical properties that are significant in three spectral band ranges. In the band range of 0.5 to 0.75 μm, for example, sensor-recorded changes may relate to leaf-pigment changes (i.e., mostly from chlorophylls). In the 0.75 to 1.35 μm (near IR) zone changes in leaf structure, primarily due to mesophyll status, may be indicated or established. The 1.35 to 2.5 μm band offers data that relate in part to leaf-water content.

Sensors operating in these ranges offer a means of surveying large geographical areas in which data are directly or indirectly related to crop stress which affects pigments, leaf, structure, and water content.

Leaf-temperature monitoring, which in turn can relate to plant-water stress caused by drought, insects, or other factors, is an application of the thermal IR scanner. Thermal IR data may be obtained during the day when terrain features generally possess temperatures higher than water or at night when the reverse is true.

When ground truth-to-sensor data correlation is established, sensor data can serve as a cost-effective tool in discriminating crops and vegetation types, in estimating crop yield, and in detecting (often before the human eye can) those areas under stress that require man's attention.

In ERTS-1 studies crop-species identification for most investigations exceeded 80% accuracy. Several investigations reported results in the 95% accuracy category (e.g., wheat identification in the Middle Western United States). The use of imagery or computer compatible tapes from more than one overpass usually increased study accuracy, but, because ERTS data are available only once every 18 days, monitoring of vegetation or crop stress is difficult. Damage assessment, however, as on dead-tree, flood, and fire zones, is possible for sites of about 10 acres or more. The ability to detect natural vegetation, forest, and crop types is related in part to the degree of spectral contrast between vegetation types being delineated.

Satellite Multispectral
Scanner Data
(.8 – 1.1 Micrometers)

Blank areas = water

```
1639
1640
1641
1642
1643                                  .AAFF+
1644                                 .EPQOMOG
1645                                 AKQRRPOOM*
1646                                BORRQSRQPM%
1647                                +KQSPPQRQOL%
1648                               .EORSRPPQRQQK.
1649                               ANQRQQQQPQOPE
1650                               .ANQRRPNQROOMD                                    DJFDEEC*                          .
1651                               .JRQNCNOQROMF                                    *KNNONONONF  ..                    .
1652                               EQPQQRRQRTQE                                    AMOOOQPQQNOQK.
1653                              *LGGOROQRROE                                    APOOQQOOPPONOOB
1654                              +LQPOONPQOK+                                    BORNPONLCCNMNQP%
1655                             BLOOOONNNOH.                                    #IONONMKJIJJJJKMCC
1656   G.                        .GPMLMNNMMNG.                                  %KPNNMMIIJJKKIJKMQO*
1657   PE                       CNNNMMMOMNMI.                                  AMRMMPMJKKJJJJIIIIOSJ
1658   ROB                      BPQNNMLMNOCKA                                 +LRMLOROKIJIJJJIJIKLROB
1659   PPK.                     IQOONMNONOOI+                                *GJMNNNMMMKHGGGIJKJKOPQG
1660   SGGE                    GRROQPOQPPNC.                               *JNKNNOPNMJJGGIGGIIKOOPNB
1661   QQQPE               *A. DOOONNNMOPPLC                              *ILLMMMNNLIJJHGHIIHILNMNA
1662   QQQQSRI*           .DKOLBHOMKLOMOMNOLG*                          .%JMKJJKLLIIKKJJJJJIIKNMMOB .
1663   RPSSSSROF+        .AEJNGFNCIPOKILLMOMLHBA.                       .AFIKKJD*%EFGKKIKKKLLJJJKLMNRQ*
1664   RQRRRSRSROH%+. .....%EKPQQRSSPGIQMIIMNNMNNK%                   .BEJKKLLKF*.+%%EKLLIGHKKLKJKJMOSRC
1665   RQRCQQRSSSSSRQQQQPPCSSRQQRSSPLHNLFFJLLLLNMLKIE*      .ABDGJIIJLMLKHA*%ECDKLIJLKIIKJKJKKMOGPB
1666   TSRPQQSTTTTSRRRRSRRRQQSSSRRQMHJOLJNPONMLOPOMMNJA    CIMLKLJJKLLMMNMLLLKJJJJLMKJIIKMMLKKKKNQQOB
1667   TSRRPQQRUSRRQQQCQRQQSRSRRQONLGILIIJNPOMKMOMMKMMGA+JNIKMLLNNNMMMLNNMKMMLKKKLLJJLJJLMNKKKKLNPRO%
1668   UTTTQRQRRTURRRRQQQQRRSQRRQPNMMQQJA*HNRRMKNCMMNMCNMLNMKNNNNOOOLMMIONMOPOOONMNMNLMLLKLMNMMMOOOQSSK%
1669   UTUUSTSSRRQSRRSSQRRRQPQOMMNNMMTSNC.EMPQNMMLKLPOOMNQMMPPOMLKKEHNOMNOOPNMMNNLLKNNNMNMNNLMNMMNNQQQQRNE*        .   .
1670   VWLSSTLLTTSRRQQQQQRFONMLKKNCKDSGI.*GJQRPMJIKJMMMMNMPPKJMNMOQPOPPNMNF*CIKKLKGIMNMMNNONONMMMLNONPQQPPPRQMHECBADHMMLC
1671   RUVWUTSSSTSRQQPPOOOMLMKLLNMJOQOLBBIKMQRSOKKKJKIJNONLIKOCOPCCKMMLMJA.,EKLLD*CKNNNNOOONMMMMMMNPPMQSQPRQSUVTSSSRSTROC
1672   QRTTUVWWUUSSRPPPOCPNMNMNKJNQQPKEGKIKLNQRSSPKKKKMNNOMIMQOKSWTSPMMMD+.*GOMJC%GOOPPQPOONONOCONOGGQQQRTTSVSTUSSSUYVRSN*
1673   RTROQRQSTSSRRPPOMNJIJIGILMNNMHEHKLIJKMPQQRGMMJIJJJKLKNTUSTSUUVSMMJBABEKONKJNMOONONMMMMNNCMLNOPMORTSTSTTSRSSSTWTSQPN*
1674   RQPQPPOPOPRQQOPRPODFQONNMJJLKGEGIIIKKMPPNQQMNMMNPMJFHNPQQSTWWWVUTNMIGIOTXWTWVQCONPSGONNMOOONMPPQQPSVUWZZYYVVVVWVTSQQOB
1675   TSRGSRSRQQRRSSTQQPGDNLJNLJIJIHJIHHLMLPQGEGHFGGLLHLQNQUTUSRUVWVXYVRTUUTWYZ111ZRMNPVWSRRQQPOPOPRQQPOORVY31YYZYWUTRRRRQQOF
1676   WSTTTSTSQORRQQQQQQQMJKNNOLIJJHHILNNNQROLG%B%.....*KTVTUUUTRTWXZZVUWXSRZZ33331VVWUVXUUYXTTQQQQRNNQONNOSZ222YXVYVQSRSTSQQK
1677   YYVSQRRCPMMPRCNNLMOPPKIPOLOOMLKLMOPQPNMLHG*.... ...*DKQQRVURUUVX1ZWXYYXTWZZWXYWY1ZXWY2YVVUWVGGPNKKMQRQNOOQUYZZYXSQUVSRTTTS
1678   UVQNQJLTVUUTSSSQQPQPQQNKNLNNMMLMPROOOOOPMHA++,*ABFLNQTWYVTUVWXYZWWXZ121ZWXZYY41ZYZZYWUTTX1VUVVVTSPPPQQSTVXYXWVRUUVVSSVXV
1679   SMCCFGNWZYWTTRQQPOOOPPPMHKMMMMNOPPOONNOMMMKBADEEGKNLLOSTSOTTVXXWWUUVUWZYVVXXXYWUY1WVRRSUY1TRTUVYXURRSRQTWWXYXRTY1YTQSRTU
1680   PJD%*AAKV11ZUUTRQGCNNPQCGLJOMOPPOPOOPONOMMNMKIKLLNONMOOLLMKQVWVSSXXUTWYWUVWWWVVUZ2YUWTRTYYURTVUUXVTVWUVVTVWXVPOTSXZUSQOQ
1681   OE+**+*CO1WXWWVSQRQOOOQPONLKOOOPOOQONOOOOPNMLMMMMNOKIKMMNLKPONRSSXWVUUUXVTWZWVUY31WXVSRVYWUTTWWUWTVWUTSRSUUWVPQTUYTSTRST
1682   RC....+DOROPPWZVUUTSRRQPCNGMOQPORQOONNOOONNMLLMMLNNMKKLMKMMLMQVXYWXYURWYWU11WUUU1211XUYXUWWUWVUXVTVVSTSTWWUVUOQVWXUTTTSS
```

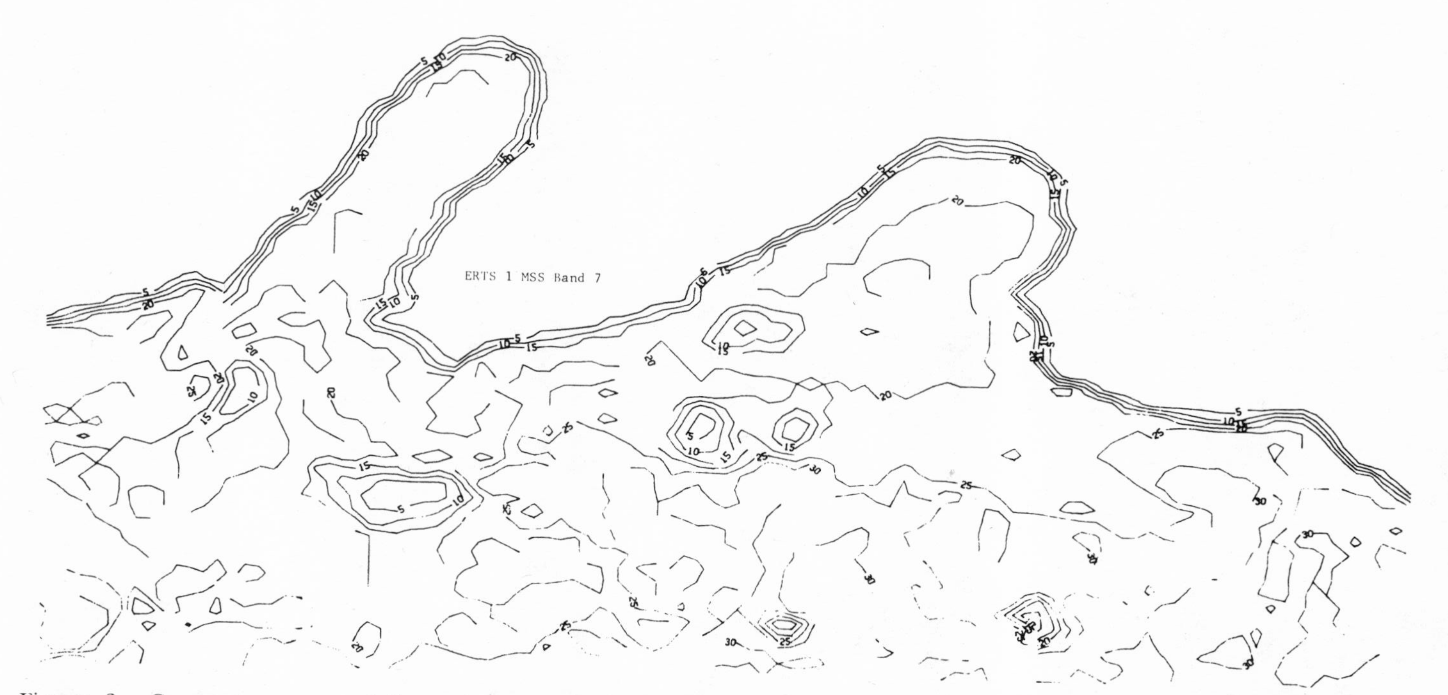

Figure 3. Computer-generated line printer map. Rio Grande region, northeastern Puerto Rico. Computer contoured from ERTS MSS 7 data.

In forestry some users of ERTS data have reported 90 to 95% accuracy in forest stratification. In the mapping of plant communities 60 to 90% accuracy figures have been established. Timber volumes have been satisfactorily estimated, and by using statistical sampling methods with ground data soil association maps were sometimes possible.

Future spacecraft systems will probably offer resolution of 30 m or better. Sensor performance specifications will be more impressive and additional spectral bands will add to the benefits (cost and technical) to the forester and farmer. Furthermore, more emphasis will be placed on computer products, and better ground truth-sampling techniques should reduce the number of ground (or control) measurements needed per satellite overpass.

Whether for agricultural or water studies or for other disciplines, most users of imagery or computer products have generally measured gray scale, color, or digital-level contrasts only. Color-additive viewers have been used with imagery to generate false color enhancements, but even these products are not quantitatively accurate because solar illumination geometry, filtration by the atmosphere, sensor error, and data-handling errors can affect measured contrast.

To obtain best quantitative results from multispectral or multiband products an optical calibration program is recommended to quantify color levels absolutely and specifically to ascertain effects in the atmosphere of scattering and absorption. The well-executed calibration program is especially important when spectral data are to be acquired on more than one date because the atmospheric scattering and absorption changes that occur necessitate the application of a unique correction to spectral data for each day acquired.

In one way of calibrating spectral data the spectral reflectance, measured optically, of a large uniform ground feature is utilized; these data then serve as the calibration standard for satellite or aircraft imagery or computer tapes. With the recording of spectral reflectance the analyst can employ any apparent spectral variation, observed from the remote sensor, to establish amounts of filtration and scattering by the earth's atmosphere. A battery-operated photometer may be used to measure spectral reflectance. Atmospheric scattered radiance can also be derived by sky photometer measurements.

Although an optical calibration program is desirable, satellite data have still proved to be useful when such programs could not be employed. For environmental studies ERTS-1 imagery and computer compatible tapes are established as tools in the survey and assessment of conditions related to water quality, land and vegetation quality, wildlife management, and the general environment. In water quality studies turbid waters, pollutant source areas, pollutant dispersion patterns, and the status of suspended solids can be detected. In monitoring land quality ERTS-1 data have been successfully employed in observance of large areas of construction, strip mines, and areas of vegetation damage caused by man or by nature. In wildlife management the types and conditions of animals and birds are sometimes related to the availability of fresh water, ground vegetation, and soil condition. Several ERTS-1 investigators reported that vegetation mapping was possible at certain sites. Also, water bodies of more than 10 acres are certainly detectable, and soil type can be inferred from vegetation existing in a given area.

In studies of the general environment satellite data have been used to observe smoke, haze, aerosols, and weather phenomena. Also mappable are fire damage, large oil spills, vegetation damage, earthquake effects, volcanos, and other short-lived events. Certainly, if data were available more frequently than every 18 days, their value would increase significantly.

In geology new linear elements, both natural and of human origin, have been detected on ERTS negatives and prints. Although no new significant mineral finds have resulted from imagery, it has pinpointed many new details related to earth-surface geology. Visible surface alterations, intersections of fractures, veins, folds, domes, and other features have been seen in satellite multispectral data.

For geologic applications in general imagery by itself can be sufficient. In the other disciplines—agriculture, water resources, the general environment, and land-use mapping, use of both imagery and computer compatible tapes is advised.

In land-use monitoring and assessment many features of 2 to 5 acre size are observable from 500-mile-high satellites. Certainly all features greater than 10 acres in size can be mapped, and high accuracy occurs in dealing with features of more than 40 acres. In interpreting ERTS data, both conventional photo-interpretation techniques and specialized computer clustering and pattern-recognition programs work well. When working with imagery, the color composite (bands 0.5 to 0.6, 0.6 to 0.7, and 0.8 to 1.1 μ), enlarged to scale 1:250,000, is a highly effective tool. Color additive viewers aid analysis operations, and may be rented or purchased in the United States from several vendors. Chapter 14 discusses specialized computer techniques related to clustering and pattern recognition. Satellite computer compatible data are also suitable for contouring, gray shading, three-dimensional plotting (radiance value = Z dimension), and color cathode ray tube display. Also, band ratioing and observing data acquired on more than one date are often best achieved by digital computer. Image interpretation generally yields satisfactory results for broad land-use classification (regional, rural), and the use of digital tapes is recommended for detailed urban or rural classification. When greater resolution is desired, satellite data products can help the user select those areas in which additional data should be acquired by aircraft and/or field check.

A major feature of the Earth Resources Technology Satellite is its data collection system (DCS), which is capable of collecting and transmitting to a receiving site the data acquired from remotely based sensors. Contained within a data collection platform (DCP), ground- or water-based transducers (usually up to eight) measure temperature, moisture, and/or other parameters. Data are transmitted to the 500-mile-high satellite, which in turn relays data to a ground receiving site. On the ground data are decoded and transmitted to a central computer. Data users can connect to this machine by a terminal. In North America DCP data can be routed by the DCS to computer and user once every 12 hours.

Because ERTS data (i.e., 185 × 185 km scenes) are available only once every 18 days at best,* the DCP/DCS approach should be considered when more frequent data are required. Also, in areas in which cloud cover is a problem (i.e., it obscures parts of images) DCP's are often ideal.

In the actual DCP shown in Figure 4 analog, serial-digital, or parallel-digital input data are acceptable. Digital words are typically eight bits. Analog transducer signals are analog-to-digitally converted and each sample is an eight-bit word. Parallel and serial data, as well as analog-to-digital conversion, are discussed in Chapter 9.

The well-executed remote sensing program, whether for water resources or agricultural or other application, generally requires, as a

*18 days with one satellite; each 9 days with two (e.g., ERTS 1 ((Landsat 1)) and Landsat 2)

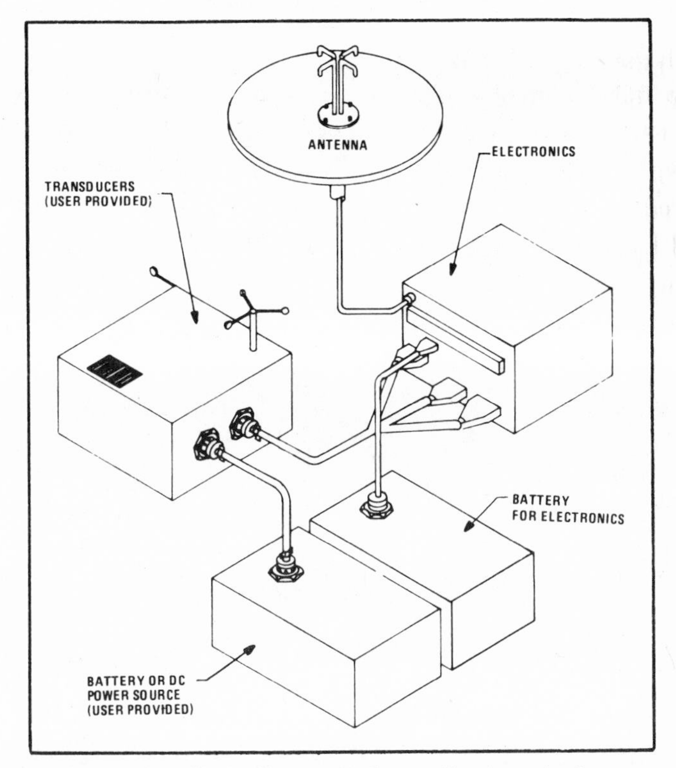

Figure 4. Block diagram—data collection platform.

prerequisite, a basic knowledge of the area to be surveyed. Knowing the types of feature and probable anomalies to be surveyed helps one to select the proper sensor bands, repetitive coverage requirements, and DCP sensors and to establish an effective data-handling program. Certainly, when DCP's are to be employed, fundamentals related to transducer data recording (analog to digital) and data processing should be mastered; these fundamentals are discussed in this book.

The comprehensive survey program also requires collection of *in situ* data for comparison with aircraft and satellite findings. The overall value of remote sensor data cannot be firmly established until such comparisons are made. When high *in situ*-to-remote sensor data correlation is achieved, sensors can be used with confidence and many expensive ground surveys can be eliminated. Ground-truth *in situ* measurements are generally made optically with a spectrometer. Spectral as well as physical, chemical, and biological ground truth can be analyzed by computer to establish degrees of correlation and to indicate the extent of intercorrelation with other variables that may not be easily observable from air- or spacecraft. When high correlation exists between remotely sensed and one or more ground-measured parameters and when statistical analysis shows that these ground factors intercorrelate with others, sensor data may be used to infer the status of other ecosystem parameters.

4

ANALOG FREQUENCY-MODULATION (FM) RECORDING OF TEST AND SURVEY DATA

Transducers used in structures testing, in aircraft flight testing, in biomedical programs, or for a host of other monitoring purposes produce output electrical signals which, by some means, must be tape recorded or relayed to analysis equipment by telemetry. This is also done with remote sensor data, collected either in an air- or spacecraft or on a remote data collection platform (DCP) that might be resting on a bridge, in a river, or elsewhere. The tape recording or telemetry technique employed must be one that generates a minimum of signal distortion.

The accepted and proved data transmission and recording technique is that of frequency modulation (FM). Even digital pulse code modulated (PCM) data (discussed in Chapter 9) must be placed on an FM carrier if aircraft-to-ground, spacecraft-to-ground, or ground-to-air transmission are to be achieved.

A familiar example of the use of frequency modulation is FM radio broadcasting, in which signals (e.g., voice or music) are detected by a transducer (the microphone), and transducer output voltage is transformed into an FM signal which is then broadcast to home radios. Frequency modulation, as opposed to amplitude modulation (AM), is not so subject to electrical interference caused by thunderstorms, auto engines, light switches, or relays. Note the difference in quality of AM versus FM on your home radio during a storm or when home appliances or light switches are used.

When critical intelligence (data) must be transmitted, that intelligence must be carried to a receiving site by an FM carrier. Why, then, are not all radio broadcasts made in FM? First, FM systems are expensive. Second, AM signals have a greater broadcast range. New York City AM broadcasts, for example, may be heard, although noisy, as far south as Virginia. FM signals, of excellent quality when demodulated, have a limited "line-of-sight" range (Figure 1). In spacecraft

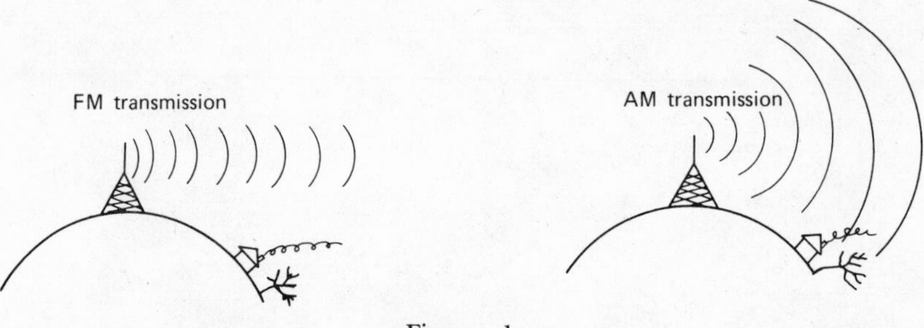

Figure 1

remote sensing acquired data are often impossible to route back to earth by telemetry because at the time of acquisition no ground receiver is within "line-of-sight" range of the spacecraft transmitter. This happened when ERTS-1 detectors were activated over South America, Europe, and other out-of-range areas. In such cases data are tape recorded, usually on a wideband recorder which has the ability to handle high-frequency data, and dumped at a later time when in range of a telemetry receiving station. (Dumped pertains to transmitting data by FM from satellite to ground.)

DEFINITION OF FREQUENCY MODULATION (FM)

A frequency-modulated signal is one in which the intelligence (data) are contained in the deviations from a carrier's center frequency. The amount that the carrier frequency varies from center is called the frequency deviation and is proportional to the magnitude of the modulating (data) signal.

Frequency-modulated signals, whether used in radio broadcasting or data recording, are characterized by their center frequency, bandwidth, and upper and lower bandedges. One example of a carrier commonly used in the aerospace industry is 25 ± 4 kHz (kHz represents thousands of cycles, or variations, per second.)

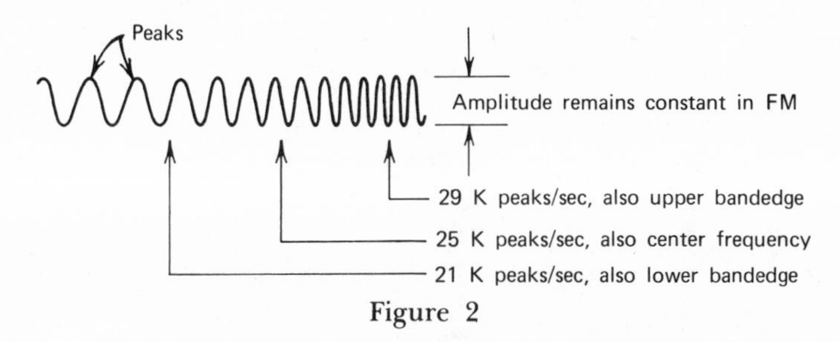

Figure 2

THE VOLTAGE-CONTROLLED OSCILLATOR (VCO)

The device that converts transducer output voltage, whether directly or following signal conditioning, is the VCO. Each VCO type performs the same basic function; that is, the generation of output signals which vary in frequency as inputs to it vary in voltage (Figure 3).

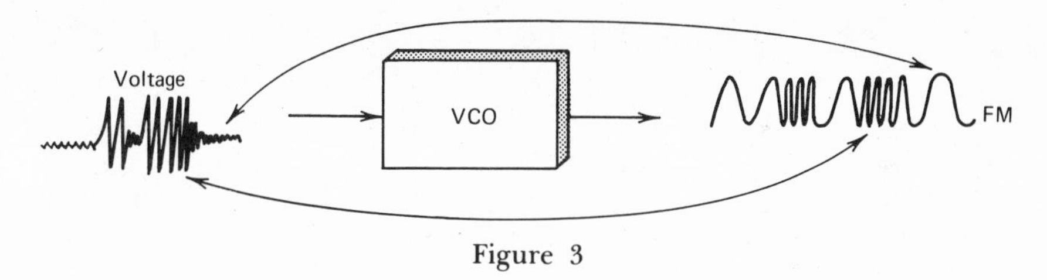

Figure 3

The VCO can be adjusted so that the higher the input voltage, the higher the output frequency.

Frequency-modulated radio broadcasting and FM data recording are remarkably similar operations, as illustrated in Figure 4. Note that in each of these illustrations a changing physical measurement (mechanical or air pressure) causes a changing transducer output voltage, which, in turn, produces an FM signal whose frequency variations are analogous to voltage variations experienced at the VCO input terminal.

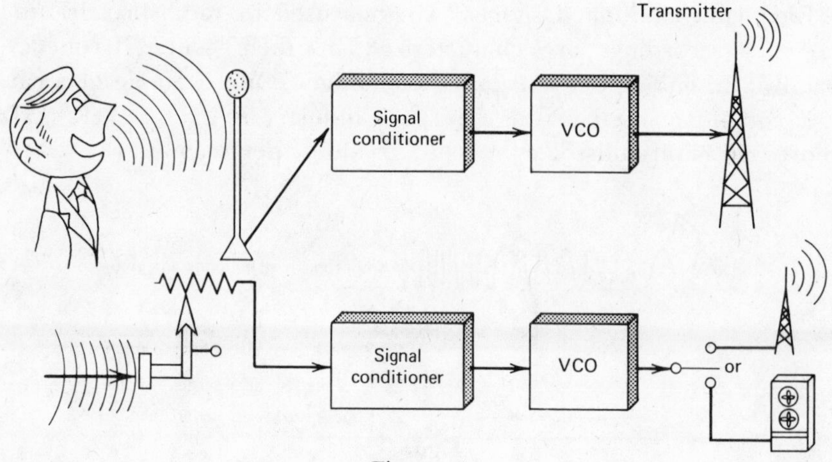

Figure 4

Because transducers generally have low-power outputs—too low to operate the VCO—signal conditioners are often required to provide the necessary power gain so that the VCO will accept the transducer output voltage.

In each of the diagrams in Figure 5 VCO's were adjusted so that the higher the physical parameter's value, the higher the VCO output frequency.

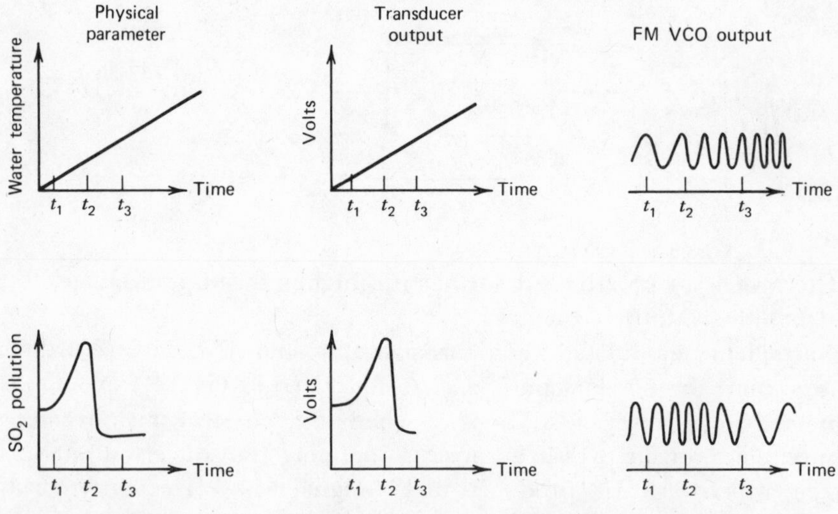

Figure 5. Additional examples of FM signal generation.

THE FM MULTIPLEX

The FM multiplex, by definition, is the signal that occurs when two or more FM signals are mixed together. Your home FM radio antenna receives what might be considered a multiplex in that many separate FM channels ("stations") are received simultaneously. You then "tune in" the channel in which you are interested.

Frequently each track of an analog tape recorder contains a multiplex. Ten carriers per track, for example, are common in certain aerospace test programs. During test data processing one may, just as in FM radio, select the channel one is interested in. A primary advantage of FM multiplexing is the ability to record many signals on each tape track. Certain FM carriers, however, mentioned in a later paragraph, are not suitable for multiplexing.

Frequency-modulated recording and the utilization of analog data reduction techniques require the recording and playback of complete and continuous data samples; 100 to 200 parameters per tape often exist. In digital recording, that is, by pulse code modulation (PCM), more than 1000 parameters can be placed on a tape. These data, however, are not complete parameter records. Instead, samples appear.

The choice of analog versus digital recording techniques must be based in part on total quantity of data to be recorded and the ability to sample high-frequency data signals accurately and economically. Also to be considered is the type of final processing to be required. When computers are required, analog data must first be analog-to-digitally converted. PCM data, on the other hand, are already converted. When analog data display are required, PCM data must be first decommutated and then digital-to-analog converted; FM data need only be demodulated.

Generally speaking, large quantities of data require PCM and the computer. Small quantities of data, especially parameters undergoing rapid change (e.g., vibration), are often FM-recorded.

When several FM carriers are to be contained in a single multiplexed signal, that signal is generated as illustrated in Figure 6. The multiplex shown may be tape recorded or subjected to another VCO for creation of FM/FM telemetry (i.e., the FM multiplex is carried on an FM telemetry channel).

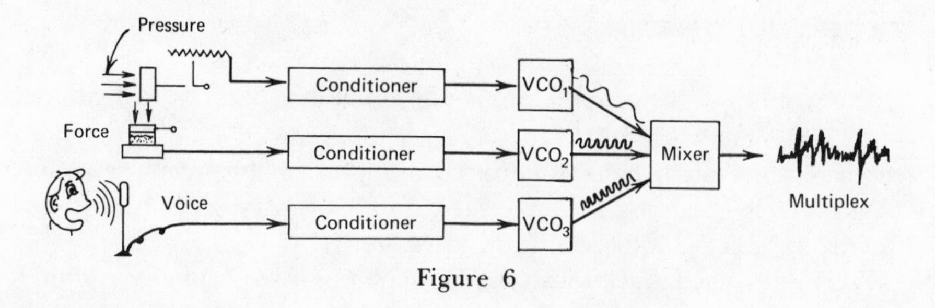

Figure 6

CALIBRATIONS

The requirement for a predata acquisition calibration for each data parameter is mandatory. A simple example illustrates this point (Figure 7).

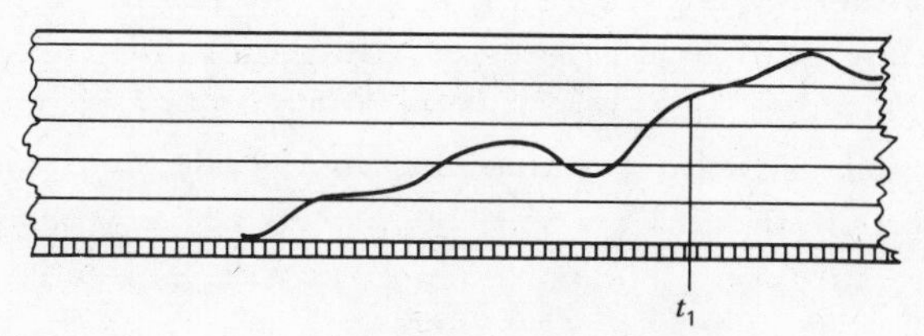

Figure 7. Time history A: temperature versus time.

Question.

What is the temperature at time t_1 for analog time?

Answer.

It is impossible to tell.

Suppose now that before actual data acquisition we record a feature known as a calibration or cal. Obviously, a cal makes data following it easily readable: t_1 is 72°F. Calibrations, when displayed with data, are the rules from which measurement values may be read (Figure 8).

A Standard Calibration Sequence

One of the standard calibration sequences commonly used by aerospace data engineers is 2½ sec of VCO center frequency followed by 2½ sec of −80% bandedge.

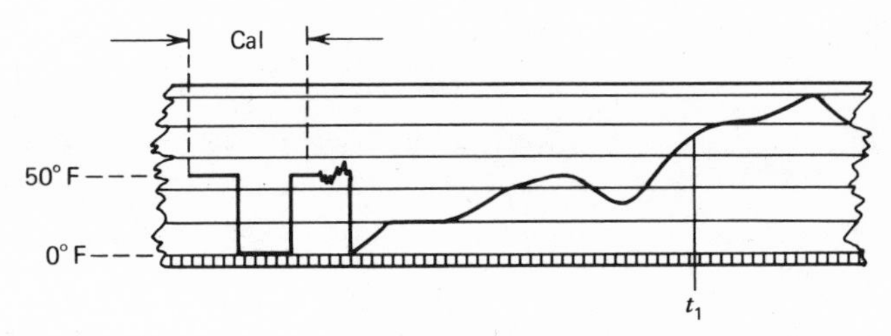

Figure 8. Time history *B*: temperature versus time.

Assume, for example, that anticipated fuel-tank pressure values during a spacecraft mission range from 0 to 40 psi. Assume also that a 25 ± 4 kHz VCO will be used to produce an FM data carrier. From Chapter 4 we know that the 25 ± 4 VCO possesses a center frequency of 25 kHz and that lower bandedge (−100% bandedge) is 21 kHz. Because −80% bandedge means 80% of the distance from center frequency to 21 kHz, this value becomes 21.8 kHz. VCO's are often adjusted so that the lowest data values produce the lowest VCO output frequency. If this is so in the above case, 0 psi can be made analogous to 21 kHz, 20 psi to 25 kHz, and 40 psi to 29 kHz. The pressure corresponding to the −80% cal (i.e., 21.8 kHz) then becomes 4 psi.

Utilization of the calibration sequence specified requires that we generate (or simulate)

2½ sec, 20 psi (@ 25 kHz)

followed immediately by

2½ sec, 4 psi (@ 21.8 kHz)

In actual practice calibration sequences are generated by simulating pressure (temperature, flow, or spectral response) by use of signal-conditioner output voltages. Simulated true cal sequences cannot be generated easily, or at all, because rapid simultaneous switching of pressure, temperature, and perhaps 5, 10, or even hundreds of measurements is usually impossible.

It must be pointed out, however, that before simulating transducer output voltages it is imperative that the acquisition engineer and/or technician know the exact relation between the input to a transducer/conditioner system and the output voltage generated. When a linear relationship fails to exist, problems can develop unless the data analyst is informed of response-curve characteristics. Often corrections can be made during data processing by the analog or digital computer.

VCO TYPES AND SELECTION CRITERIA

In designing a data acquisition plan, care must be taken to choose proper VCO's. Similarly, data-handling personnel must be aware of VCO characteristics so that data processing can be performed with hardware that matches the frequency-response characteristics of each VCO-generated FM carrier.

Hundreds of different VCO's exist. Among common types are AIA, Constant Bandwidth, IRIG, Wideband Group 1, and Wideband Group 2. One should never choose a VCO whose frequency response (usually expressed in hertz) is less than the highest expected data change rate.

If, for example, VCO 25 kHz with frequency response = 2 kHz were available but rocket vibrations (5-3000 Hz) were to be measured, this VCO would be unacceptable. If, on the other hand, we were to record human-pulse information during a medical test, that is, approximately 74 pulses/min (normal) to approximately 100 pulses/min, then VCO 730 ± 55 Hz which has a frequency response of 11 variations/sec (or 660/min) would be adequate.

Selection of a VCO with a frequency response that is too small introduces amplitude errors in final processed data. Selection of a VCO with a frequency response considerably higher in value than that of the maximum expected data frequency is also unwise: (1) several types of high-frequency response VCO cannot be multiplexed; (2) high-frequency response VCO's require that high-frequency response (and high-cost) tape recorders be used. Among the types of VCO available are the following.

1. *Constant Bandwidth (CBW).* A series of carriers, each of which has a different center frequency but possesses a constant bandwidth. One series is 25 ± 4 kHz, 40 ± 4 kHz, 55 ± 4 kHz, 70 ± 4 kHz, and 85 ± 4 kHz.

2. *Wideband (WB).* A series of carriers, each of which has the same center frequency but a bandwidth that is extremely wide. An example is 108 ± 43.4 kHz. Although wideband carriers possess high-frequency responses, they may not be multiplexed.

3. *Proportional Bandwidth (PBW).* A series of carriers, each of which has a different center frequency but possesses a bandwidth that is a percentage of the center frequency; PBW carriers possess

bandwidths that are considerably narrower than WB and may be multiplexed.

Example. 400 Hz ± 7½% = 400 ± 30 Hz (IRIG Band 1)
540 Hz ± 7½% = 540 ± 42 Hz (IRIG Band 2)

. .

. .

. .

70,000 Hz ± 7½% = 70,000 ± 5250 Hz (IRIG 18)

Generally, IRIG ± 7½% packages contain approximately 18 VCO's. Wideband (±40%) has two prime advantages over IRIG proportional bandwidth and constant bandwidth packages:

(a) Higher frequency response.
(b) Unlikely to be affected by tape speed errors. (Tape speed error problems are discussed in Chapter 6.)

DEFINITION OF FREQUENCY RESPONSE

$$\text{FR} = \frac{\text{deviation in hertz from center frequency to bandedge}}{\text{VCO modulation index}}$$

The modulation index is a number that physically represents each type of VCO. The lower the index, the higher the system frequency response. "Modulation index" is sometimes referred to as an "engineering design number." For IRIG channels modulation index = 5. For CBW channels modulation index = 2. Wideband VCO characteristics offer an index close to that of CBW.

5

MAGNETIC DATA TAPE RECORDERS

Tape recording of *in situ* transducer and remote sensor data occurs frequently. Air- or spacecraft undergoing flight test, for example, record data with an onboard tape recorder and/or telemeter data to a ground station for recording and processing. Often, for safety of data preservation, data are both FM telemetered to the ground and either analog or digitally recorded in the vehicle. If one system fails, backup data are still available.

In satellite remote sensing of earth resources multiband data from the multispectral scanner are sometimes acquired over a region not in range of a receiving station. Here data are tape recorded for later satellite-to-ground transmission when a receiving station comes into range.

Generally speaking, tape recording is essential when a copy of original data is to be preserved, when data transmission at the time of acquisition is impossible, or when data processing hardware cannot handle all transducer or sensor outputs simultaneously. Depending on the quantity and type of data to be recorded, either a ¼-, ½-, or 1-in. tape-width recorder may be selected. In ocean meters, when only temperature, pressure, conductivity, and current speed and direction are to be recorded, a low-price ½-in. recorder may be used. In vehicle testing, when vibrations, pressure changes, strain, flow, and other parameters must be monitored, sophisticated tape recording systems (½ or 1 in.) costing in excess of $50,000 are often required.

Basics of magnetic data tape recording are discussed in the paragraphs that follow. Selection of a recorder (and its price) are generally dependent on the desired recording capacity, frequency response, and recording/reproducing accuracy specifications. The construction of a magnetic data tape is shown in Figure 1.

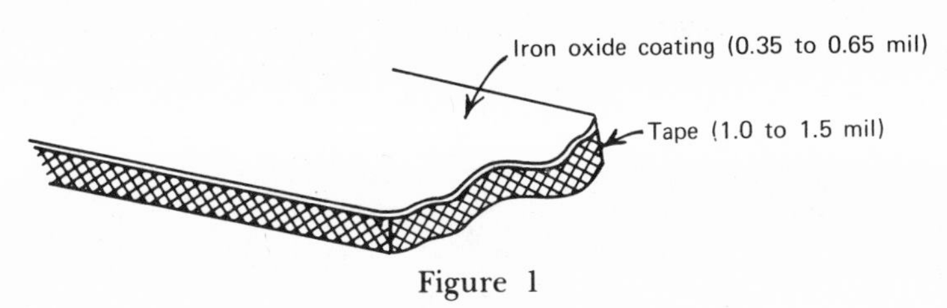

Figure 1

FUNDAMENTALS OF RECORDING AND REPRODUCING

Recorded data tapes, whether one that contains continuous analog signals or digital pulse data, are treated with a coating that has been magnetized. The device that impresses data onto this coating is an electromagnet (Figure 2). As tape reels turn and tape passes across

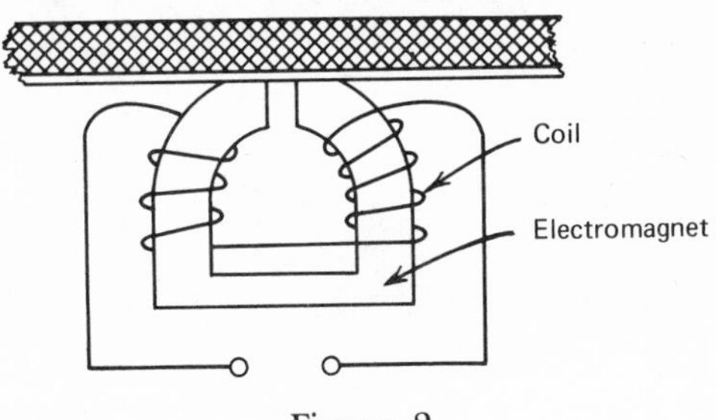

Figure 2

the electromagnetic gap, information converted to variations in electric current is passed through the coil around the ring structure. The magnetic field that results is concentrated in the ring structure and produces a field that fringes at the gap. Information to be recorded (i.e., originally from the transducer or sensor) is converted to variations in flux intensity at the gap. These variations are then transferred to the magnetic film on the tape as it is made to move past the tape-recorder record head.

In reproduction, or tape playback, the gap scans the tape and responds to the changes in its magnetic intensity. Voltage is induced in the head coils whenever changes in magnetic flux are sensed.

In comparing one tape transport with another, one usually describes the system in terms of the number of tracks, tape width, recorder frequency response, dB (power) variation, tape speed, and bandwidth. As an example, consider a system specified as follows:

- 1 in.-14 track.
- Frequency response = ±3 dB.
- Bandwidth = 300 to 750,000 Hz.
- Speed = 60 in./sec.

These specifications indicate that although tape runs at 60 in./sec passed reproduce heads the amplitude gain or loss of a recorded signal on any of the 14 tracks will not vary more than ±3 dB from an established reference. Signals will fall within the ±3 dB window no matter what their frequency; either 300 Hz, 750,000 Hz, or anywhere in between.

Assuming that the established input reference is 1 V, the output following tape recording and playback will not vary from the 1-V level by any more than ±3 dB (or, in this case, approximately 30%) (Figure 3). *Tape recorder output may vary from the 1 V level by ±3 dB to be in "spec."*

A 3-dB error in analog FM or digital recording is not significant because FM demodulation systems deal with frequency of the re-

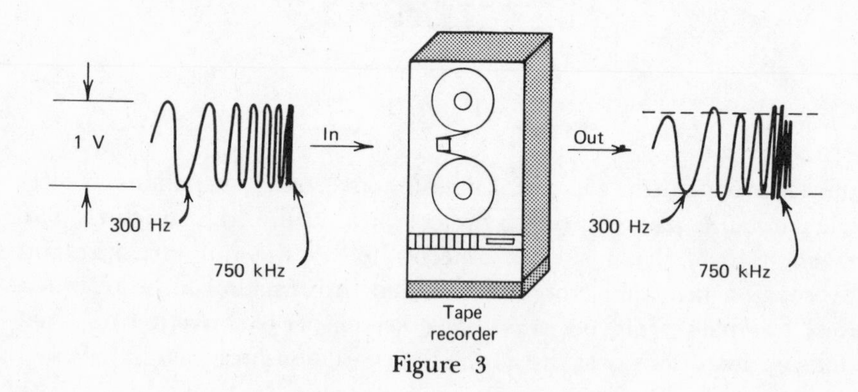

Figure 3

corded signal—not its amplitude—and digital devices look for the presence or absence of a pulse and are not designed to reject outputs if small amplitude variations (in a pulse) exist.

High-quality tape recorders, like the expensive race car, require critical adjustments. A poorly tuned transport can cause output data to suffer from the effects of, among other things, harmonic distortion, nonlinear phase response, wow and flutter, skew, and jitter. Each of these common error sources is discussed briefly.

Harmonic Distortion

The production of undesirable components which are integer multiples of the original data frequency (Figure 4).

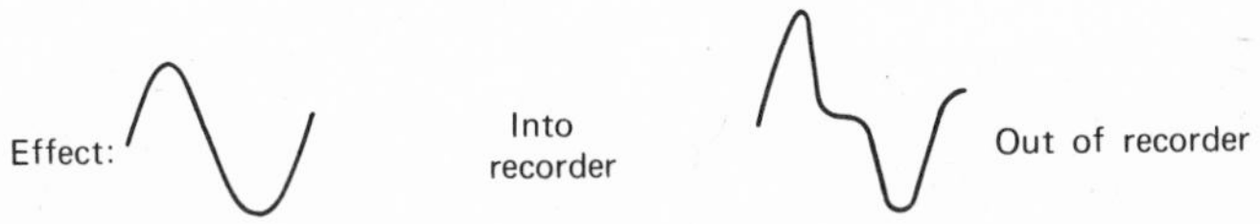

Effect: Into recorder Out of recorder

Figure 4

Nonlinear Phase Response

Tape recorders can cause phase-shift errors unless adjustments are made to form a linearly increasing phase shift with increasing frequency (Figure 5).

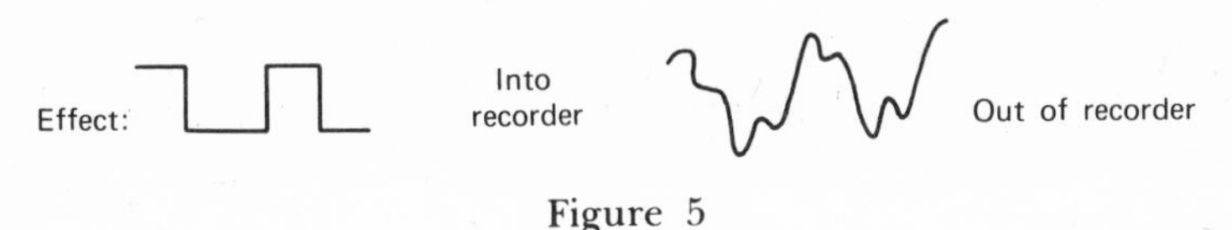

Figure 5

Wow and Flutter

Nonuniform variations in tape speed produce an increase in the true data frequency if the tape runs too fast and a decrease if the tape runs too slow. Since *all* tape recorders experience speed variations, corrections for wow and flutter may be essential. (Wow pertains to speed variations that are slower than the faster flutter variations). Chapter 6 discusses a typical analog correction system related to FM data.

Skew

Skew, or the intertrack displacement as the tape passes over transport heads, may also be thought of as a change of azimuth as the tape-to-head contact is made. Here tape motion causes an angular velocity between the gap center line and a line perpendicular to the tape center line. Normally expressed in microseconds, skew detects the relative time base correlation differences between tracks.

Jitter

A time-displacement error caused by the displacement of a point on tape from where it should have been. Jitter is dependent on the frequency composition of flutter.

In addition, problems may develop if tape or tape-reel quality is poor. Sources of error include loose oxide, dust or foreign matter on the tape, tape scratches or creases, misaligned reels, and varying oxide thickness. Another source of error is improper tape handling and storage. A dropped tape, unexercised tape, or one subjected to harmful environmental conditions during storage or shipment can suffer serious degradation.

Details pertaining to proper tape-storage and -handling procedures are usually available from major producers of magnetic data tapes. Several offer seminars on this not-to-be-ignored topic.

6

FM DATA DEMODULATION

Test data from an aircraft flight test, when telemetered to a ground station, are carried on FM. Spacecraft, equipped with multiband remote sensors, similarly employ FM to route raw data to ground receiving sites. In FM radio broadcasting each station transmits a unique FM carrier to the home receiver or FM radio.

The similarity between the manner in which the home radio acts on receipt of FM and the way in which the data processing facility FM data discriminator treats FM telemetry or tapes which contain FM recorded data is remarkable.

The home FM radio antenna receives what might be considered a "multiplex", or more than one FM station's, signal. Because each station (per geographical region) employs a separate voltage control oscillator (VCO) in FM signal generation, we may select a station (channel) simply by tuning in its VCO center frequency (Figure 1).

At the telemetry receiving station or in the tape-handling processing facility channel selection is achieved with an FM data discriminator. This unit is similar to an FM radio in that it contains a tuning system and an output-strength adjustment control. In actual practice tuning is achieved by plug-in modules or a variable frequency bandpass filter through which individual channels may be selected (Figure 2).

By taking a closer look at the FM data discriminator (Figure 3) we may observe its four basic sections: the

1. *Tuning unit* (plug-in or dial-in). Bandpasses center frequency ±

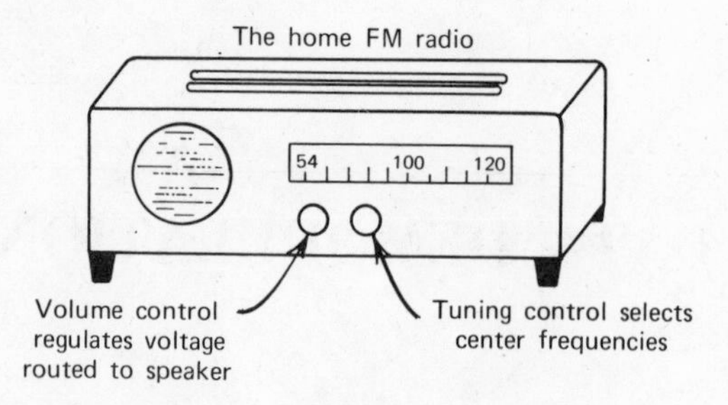

Figure 1

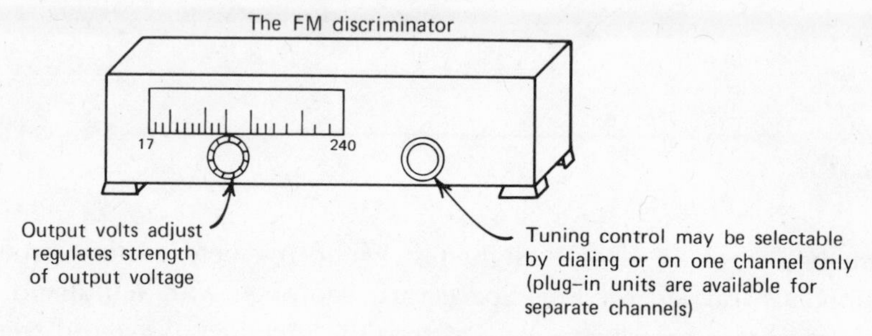

Figure 2

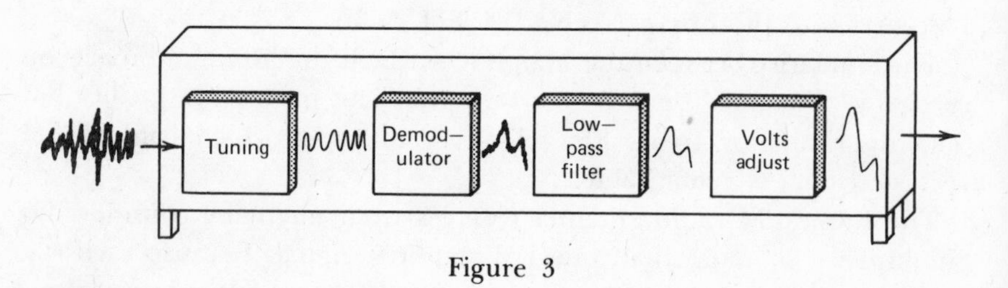

Figure 3

bandedge and sharply attenuates or rejects all other frequency components.

2. *Demodulator.* Converts the bandpassed FM carrier into a voltage with variations analogous to carrier frequency variation.

3. *Low-pass filter.* Sharply attenuates or rejects high-frequency noise components above the highest expected data frequency.

4. *Output voltage adjust.* Similar to the volume control in an FM radio, but instead regulates the strength of the output voltage. Outputs may be routed to analog analysis and display hardware or be analog-to-digitally converted for computer entry.

After demodulation of several FM carriers from a tape recorder, for example, data may be directly routed to a scope, an oscillograph, or a pen recorder for display (Figure 4).

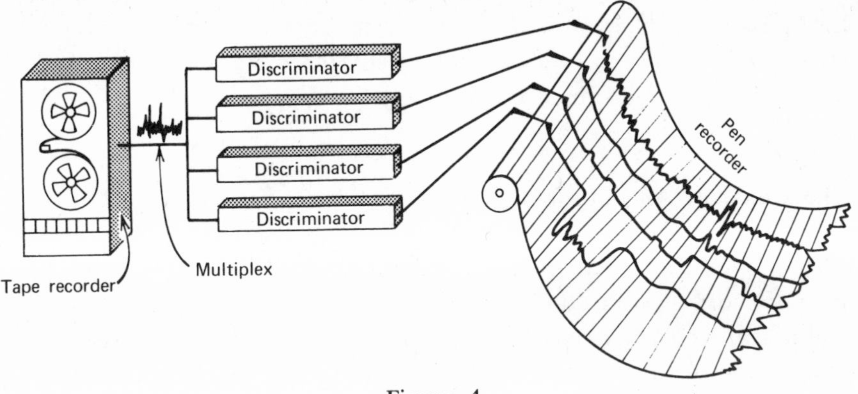

Figure 4

LOW-PASS FILTERING

The low-pass filter, present in most demodulation systems, is designed to eliminate high-frequency noise components from processed data (Figure 5). Extraneous or undesired high-frequency components are

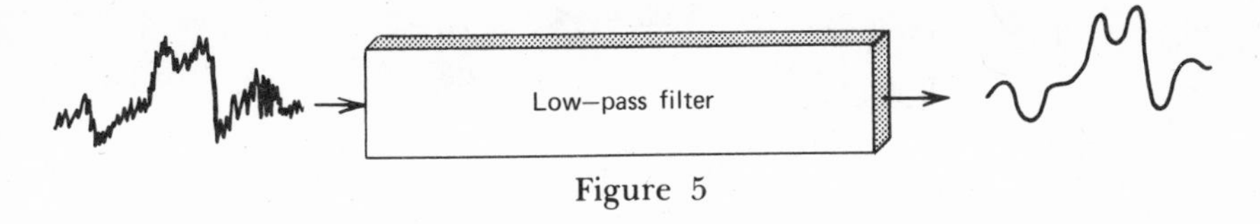

Figure 5

sharply attenuated or eliminated. Only the lower frequency data components are passed through the filter.

As stated in Chapter 4, each VCO has a unique frequency-response rating. Frequency response relates to the maximum number of data variations per second that may accurately convert to FM by the VCO. If during FM recording data rates smaller than or equal to the frequency response (FR) were common, we should expect FM demodulated data to show data change rates smaller than or equal to the (FR) rating of the original VCO. Here, a low-pass filter rated at (FR) would sharply attenuate all data with variations greater than (FR). These variations are often due to electrical interference, poor hardward grounding, or "noisy" electrical circuit components. When a VCO's (FR) is 200 Hz but the maximum number of data variations per second is 100, a low-pass filter of value ½(FR) is often desirable. Low-pass filter selection is a bit tricky; selection criteria must be mastered, and in most cases a choice between the constant amplitude and constant delay types must be made.

The Constant Amplitude Low-Pass Filter

Also commonly called the Butterworth filter, the constant amplitude filter is often utilized in data processing operations in which amplitude accuracy is more important than phase. A salient feature is that of maximally flat amplitude response.

Although the constant amplitude filter produces negligible output-voltage amplitude errors, it is generally unsuitable for operating on data that contain "step-type" transitions.

- When a data signal is continuous and without "step" changes, the constant amplitude filter can effectively remove high-frequency components and pass, with excellent amplitude accuracy, low-frequency intelligence (Figure 6).

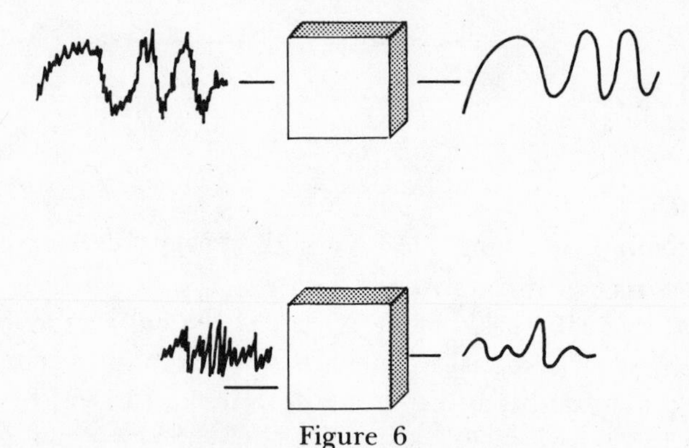

Figure 6

- With "step" data, amplitude errors are generated after each abrupt data transition (Figure 7).

The Bessel or constant delay filter should be used for step data. Its

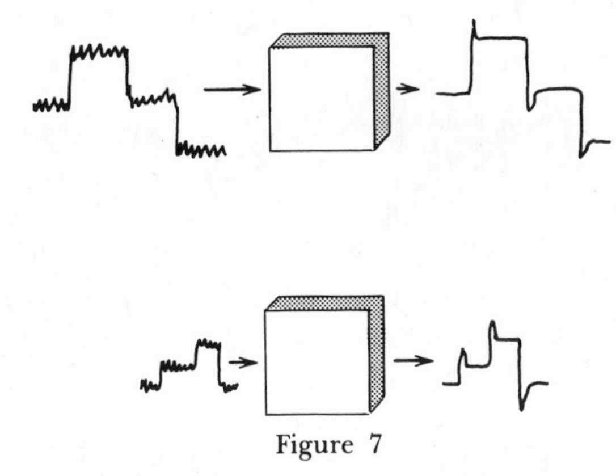

Figure 7

salient feature is fast-settling time and lack of data overshoot. It is especially ideal for reconstructing sampled waveforms. The Bessel filter, unlike the Butterworth, will not allow the 10 to 15% overshoot (to a step function). Overshoot instead is held to 1 to 2%. Additionally, the Bessel can prevent jitter by virtue of its ability to provide fixed time delay. Although its roll-off curve is similar to the Butterworth, it cannot attenuate high-frequency components as well.

- Response characteristics, Butterworth and Bessel (Figure 8).

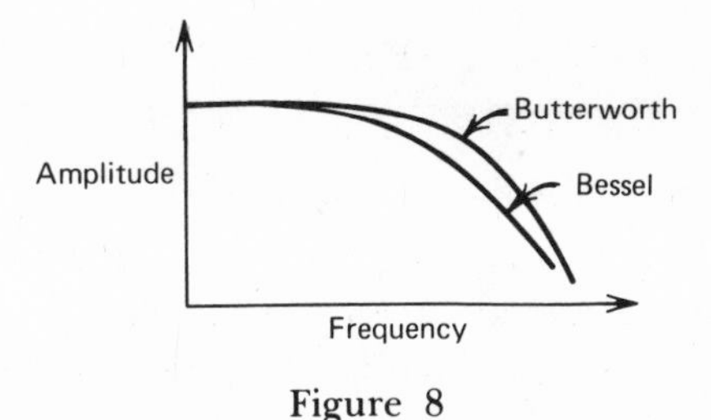

Figure 8

Roll-Off Curves

If we placed a constant amplitude filter of value f_u at an FM discriminator's output terminal and if f_u is the upper data frequency of interest, we should find that the Butterworth filter's output would remain relatively constant in amplitude except for data with frequencies near or at value f_u (Figure 9).

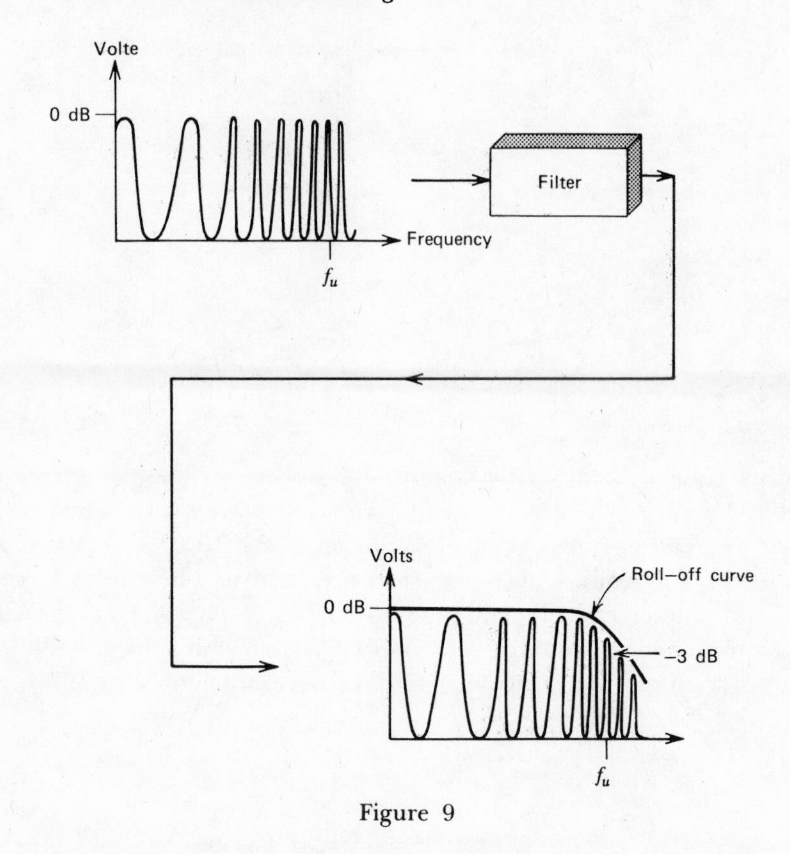

Figure 9

For many low-pass filters the amplitude attenuation at f_u is -3dB, but all data frequency components above f_u are either totally eliminated or sharply attenuated. All low-pass filters have unique roll-off curves; no single filter can totally reject all components above f_u or precisely preserve those below f_u.

Among other commonly used filter types are the Chebyshev and the elliptic (or Cauer). The Chebyshev allows greater frequency selectivity than the Butterworth or Bessel but offers a roll-off curve with ripple. The elliptic provides maximum attenuation slope after f_u and is extremely useful in telemetry or sonar data filtering, that is, when sharp attenuation between each frequency (intelligence) band is desirable.

- Response characteristics, Chebyshev and Elliptic (Figure 10).

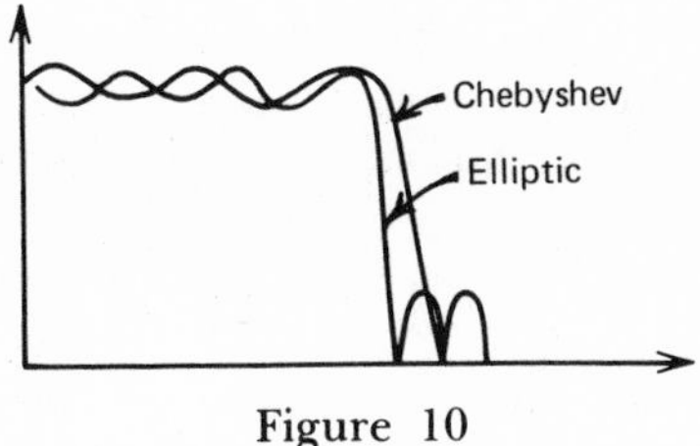

Figure 10

Low-Pass Filter Selection Criteria

Selection of a low-pass filter for a discriminator or an electrical signal from another source should be based on several key factors, stated in the form of these questions:

1. What is the highest expected frequency value of meaningful data?
2. Are data of the continuous or "step" type?
3. Will roll-off characteristics of available filters introduce harmful amplitude or frequency error?
4. What is the frequency response of the FM data carrier?

Also, of the two available design types of filter, the *active* analog filter can perform the bandpass (or band reject) operation better than the *passive* filter in frequency ranges below 100 Hz. The passive filter is generally more expensive and passive inductors lack temperature and component-value stability at 100 Hz and below. Above 100 Hz, however, the passive low-pass filter proves to be ideal.

EFFECT OF TAPE-SPEED ERRORS ON DISCRIMINATOR OUTPUTS

When tape recorders run too fast, tape-reproduce heads see FM carriers appearing higher in frequency than the signal originally recorded. The reverse is true when tape passes the reproduce head more slowly than normal.

Because FM discriminators are frequency-to-voltage converters, tape-speed errors lead to voltage errors at discriminator output stages.

The following example illustrates the effect on final analog displayed data when a recorder runs too fast. Discriminator calibration here is that center frequency in causes 0 V out and upper bandedge in causes +10 V out.

1. Normally, if 25 kHz were recorded originally and the tape playback unit ran on speed, 0 V would appear at the discriminator output (Figure 11).

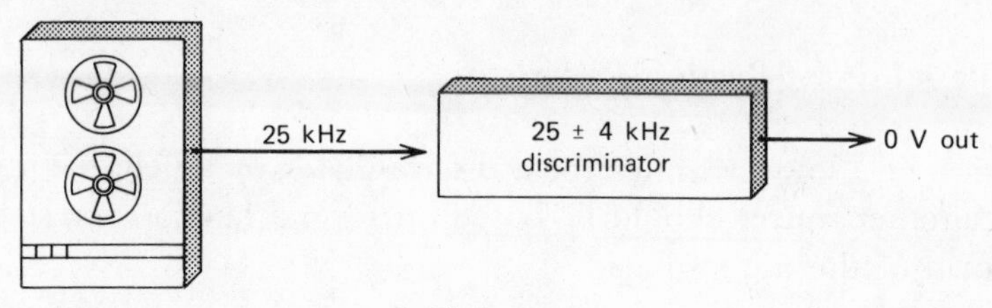

Figure 11

2. Now, if 25 kHz were originally recorded but playback speed was off (i.e., by 0.4%), a +¼-V error would occur (Figure 12). Assume,

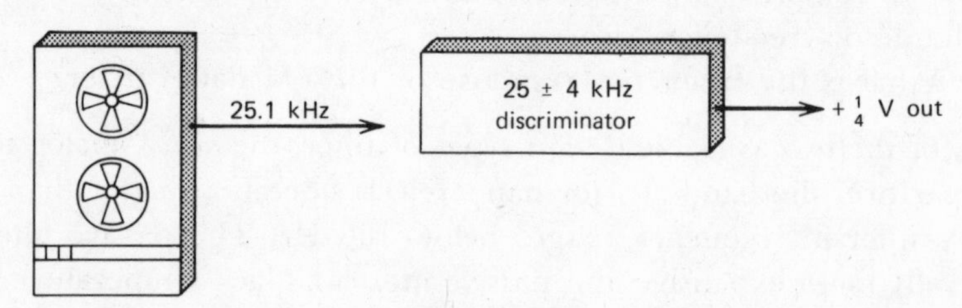

Figure 12

too, that the measurement involved was temperature and that the 25 ± 4 kHz band represented 100 ± 100° F, or a span betwween 0 and 200°F. When discriminator output data is routed to a pen recorder, or oscillograph, note the effect of a tape that is running too fast (Figure 13).

Obviously small tape speed errors can cause significant amplitude errors in final data. This effect is even more disastrous with narrow-band FM carriers (e.g., IRIG channel 1: 400 ± 30 Hz) than with wideband carriers (e.g., 108 ± 43.4 kHz).

One method of preventing voltage errors as tapes run too fast or too slow is wow and flutter compensation or, as it is commonly called, *tape speed compensation*.

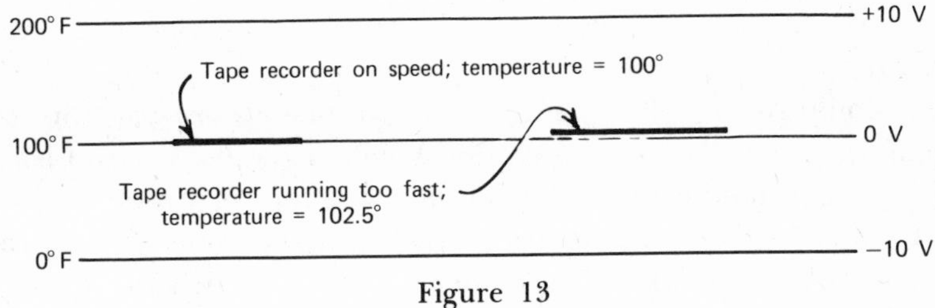

Figure 13

Use of this technique first requires that during original data tape recording a reference tone (compensation frequency) be added to the typical FM multiplex. Later, during data processing, a compensation discriminator and error-detection system makes error corrections (Figure 14). This tape speed compensation system is recommended when

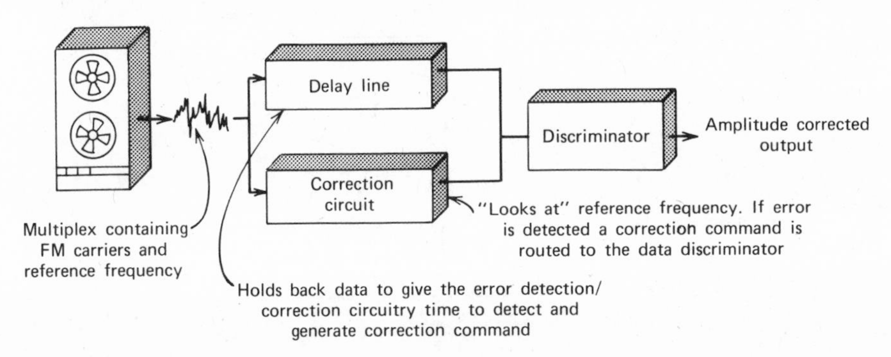

Figure 14

nonwideband VCO's are to be used and/or when corrections for amplitude are mandatory. It must be remembered that tape speed errors of ±(¼ to ½)% are not uncommon.

VOLTAGE-CONTROLLED OSCILLATOR FREQUENCY DRIFT

A common problem in FM data recording appears when temperature or other environmental factors cause the output frequency from a VCO to drift upward or downward.

Processing such data through a discriminator causes voltage error in final outputs. Error correction is possible, however, if it is recorded with data subcarriers and the reference tone is a signal commonly called the *command band*.

The command band activates an *auto-cal* or *scale-and-zero* correction system to shift drifted data (i.e., discriminator output voltage) back to 0 V during a standard calibration. Following the correction of subcarrier calibrations, data signals are corrected by the same factor, namely $+X$ or $-Y$ V, depending on whether a calibration drift of $+X$ or $-Y$ V has occurred.

A typical display of a data calibration waveform and the command signal is made in Figure 15. The command band differs from the

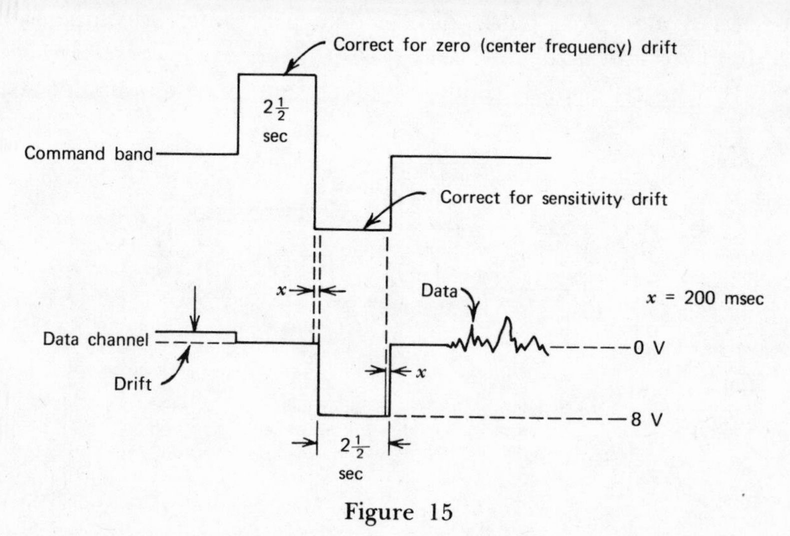

Figure 15

reference frequency in that the command band is used to make corrections caused by environmental or electrical factors; the reference frequency permits compensation for errors caused by tapes running too fast or too slow. VCO drift is a common problem in aircraft data acquisition in which environmental factors vary in different sections of the flight path.

FREQUENCY TRANSLATION

To utilize the full bandwidth capability of the modern tape recorder effectively data acquisition personnel must ensure that a maximum number of data channels are multiplexed for entry onto each tape track. One means of accomplishing this objective is by the process of frequency translation.

An example of a frequency translation technique commonly employed, for example, in air- and spacecraft vibration testing, is illustrated in Figure 16. Note that frequency translation effectively

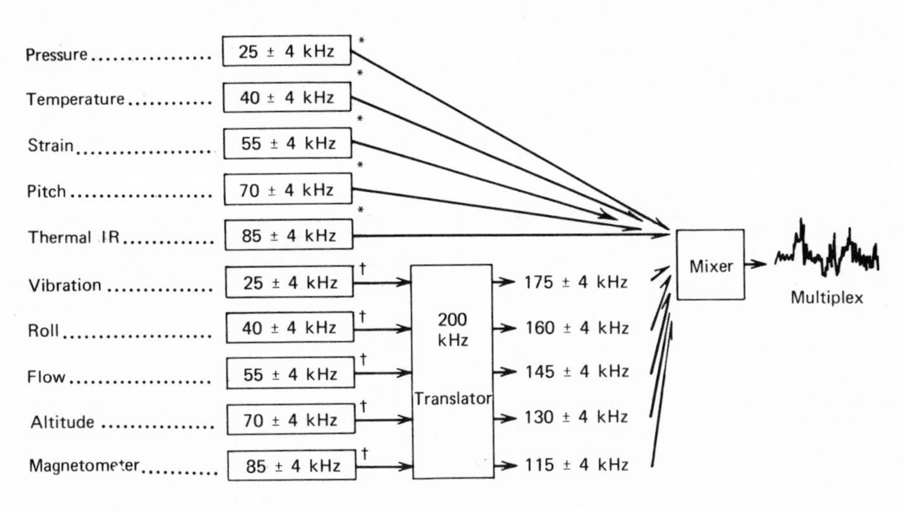

Figure 16

doubles the number of channels that may be recorded on each tape track. Although this example contains two sets of identical VCO's, the end result is the creation of one set of subcarriers (channels) that may be directed to one tape track. None of the 10 will interfere with one another as pressure (25 kHz) and vibration (25 kHz) would if they were presented to the same tape track. Just as FM radio stations use different broadcast center frequencies, so must we in recording FM carriers onto a single tape track. Sum and difference frequencies are created in the translator, and sums may be rejected because they require a higher frequency response tape recorder than the differences. (Higher frequency response recorders are generally more expensive.)

Data playback is the reverse of the above. A detranslator is employed, and two sets of discriminators (with values equal to the original sets of VCO's) serve to demodulate each subcarrier.

7

TIME CODE

Time code is a precisely coded analog or digital signal that may be tape recorded or telemetered simultaneously, with all data detected by transducers and remote sensors.

When time code is processed or displayed with data, the data analyst is provided with a highly accurate reference signal from which the time length of each data event can be measured and a means of specifying where on a magnetic data tape or where in a stream of data an anomaly or significant event has occurred.

In flight testing aircraft (where tapes are recorded) or in recording thermal line scanner data, flight logs can contain a time column into which the start and stop times of maneuvers or each thermal scene are recorded.

In the data processing laboratory, in which aircraft tapes undergo processing, the machine operator performs data reduction by time zones. Instead of machine processing an entire tape, he "zeros in" only on time slices of interest. Time code thus offers a means of reducing a total tape more efficiently, and a method of decreasing both labor and machine time costs by acting only on tape regions of interest to the data user.

Figure 1

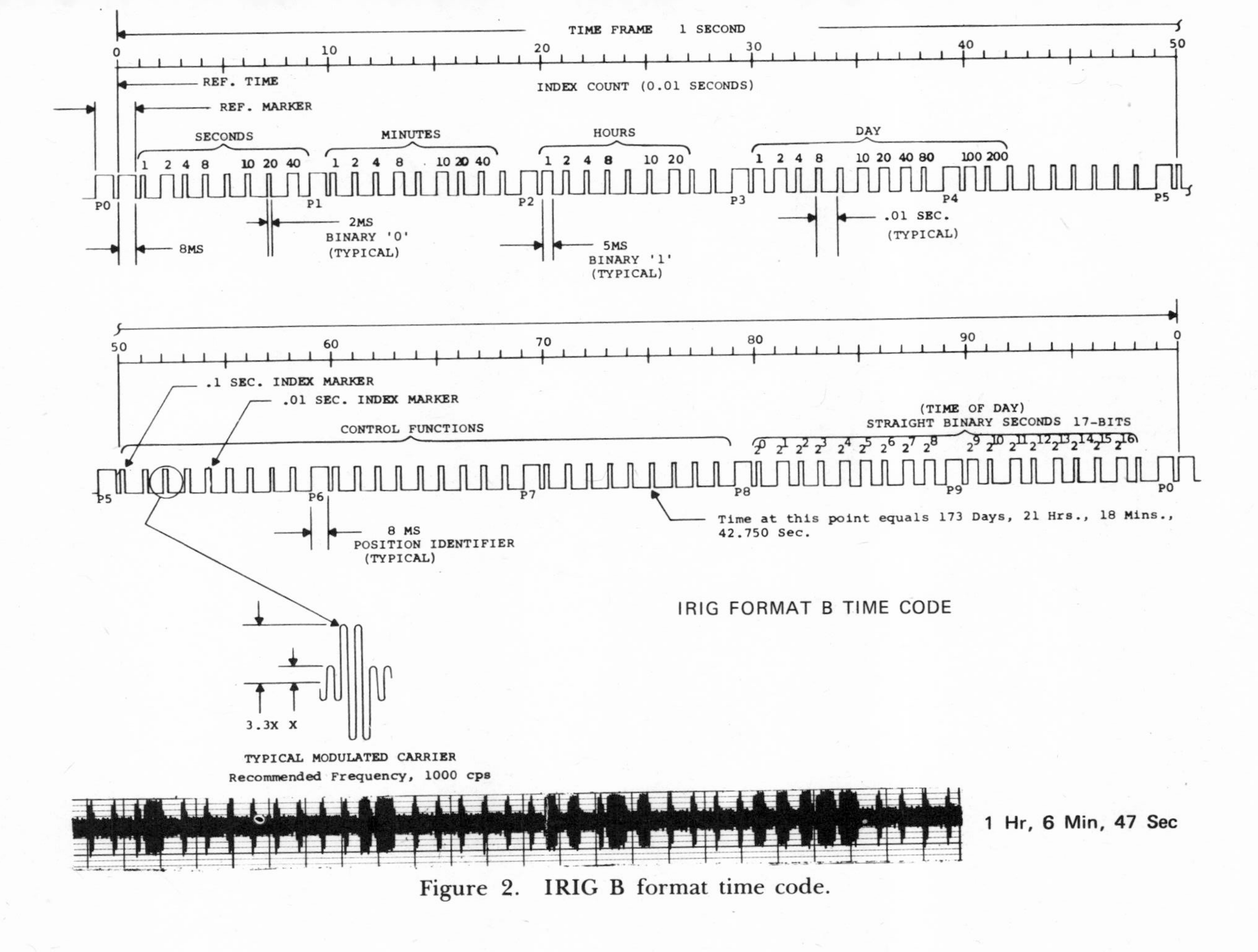

Figure 2. IRIG B format time code.

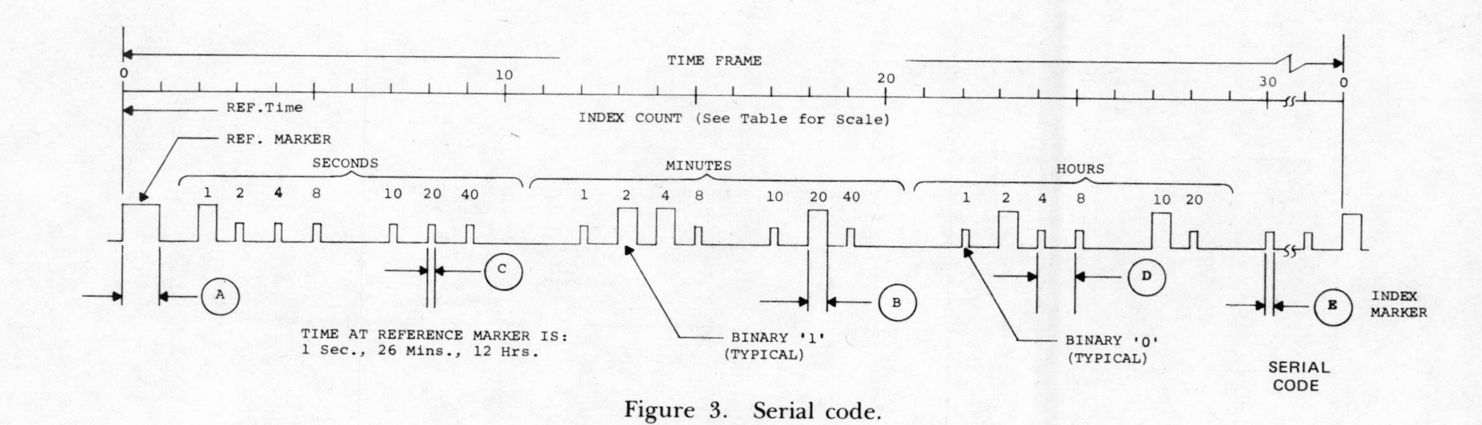

Figure 3. Serial code.

More than 10 different time-code types are in existence: IRIG,* NASA, Elgin, and Special Serial are examples. Each code is generated by a device called a time-code generator and can be read and displayed by a time-code translator. Both devices usually contain nixie tubes that provide a lighted display of actual generator or translator-seen time. Each nixie tube can display a "1," "2," or other character ("3" to "9"), including "0."

The commonly utilized IRIG B code is shown in Figure 1 in its analog waveform. Some translator systems are capable of generating an easy-to-read and -display waveform known as slow code. It is suitable for routing, along with data, to the chart recorder or the oscillograph.

IRIG B format specifications follow (Figures 2 and 3). Also shown is an example of the 1-sec slow code (Figure 4). Slow code is usually the

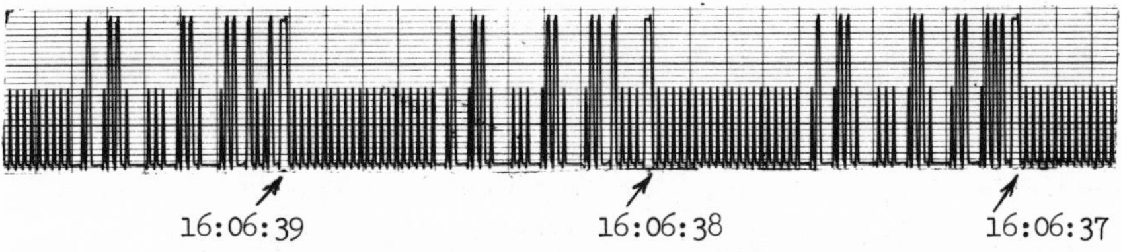

Figure 4. One-second serial code.

1-sec, 10-sec, or 1-min type; time updates are fed into a display device once each second, every 10, or once each minute, respectively.

*Inter Range Instrumentation Group.

8

ANALOG DATA DISPLAY TECHNIQUES

The modern data processing facility offers a large variety of display devices for analog and digitally processed data. Analog data are usually suitable for direct connection to the input terminals of these devices, whereas digital data (i.e., from a computer) must be subjected first to digital-to-analog conversion before the display operation can commence.

Analog display devices are commonly referred to as *chart recorders.* Among the recorder types commercially available are the following:

1. Pen or heated stylus strip chart recorders.
2. Galvanometer oscillograph recorders.
3. Fiber optics oscillographs.
4. *X-Y* and *X-Y-Y′* plotters.

Also useful for several applications is the oscilloscope. When equipped with a storage cathode ray tube that retains images, oscilloscope data may be photographed by a Polaroid or other camera.

In selecting the proper analog display device for the laboratory or data processing facility, it is advisable to lean toward the potentiometer-type chart recorder which produces displays by the capillary ink or electric-writing technique. Potentiometer recorders provide large chart display widths, excellent resolution, and low system drift. Charts generated by this device are almost always totally suitable for reproduction on any standard office copying machine. A major limitation, however, with the potentiometer recorder is that of frequency response. When original data contains variations of more than 50/sec, for example, another recorder type must be used if amplitude accuracy is to be preserved.

When data frequency rates are high (i.e., up to 5 kHz), the light beam galvanometer recorder should be used. The subminiature-type galvanometer contains a lightweight mirror that moves (twists) proportionately to variations in the electrical signal to which it is connected. As mirror movements occur, a light beam (directed at photosensitive paper) moves and the data are recorded. O'graph paper, as it is commonly called, fades with time unless chemically treated. Some paper records are rated at only 1000 hr in normal room light. Direct sunlight can easily destroy all images recorded.

To record extremely high-speed data to about 1 MHz in frequency value one must turn to the fiber optics oscillograph which transmits cathode ray tube images through a fiber optics bundle and onto photosensitive chart paper or film. One system, the Honeywell Visicorder, was used to produce the thermal infrared imagery described in Chapter 2.

The basic principle behind the operation of the galvanometer-type recorder is based on a coil carrying an electrical signal. When placed in a magnetic field, the coil will rotate, depending on the magnitude of the signal. Motors, ammeters, and voltmeters are examples of devices operating on this principle. So, too, is the pen recorder illustrated in Figure 1. Being heavier than a small mirror, the pen is generally limited in its physical ability to oscillate accurately at frequencies above 60 to 100 Hz. Lightweight mirrors, on the other hand, can oscillate accurately several thousand times each second.

Contrasting the pen recorder, the mirror-type galvanometer recorder, and the fiber optics oscillograph, we obtain the following:

Pen recorder. High accuracy, attractive displays, but extremely limited in frequency response.

Galvanometer oscillograph. Frequency response to 5 to 10 kHz but poor quality display because of trace width and photographic type of recording.

Fiber optics recorder. Exceptionally high frequency response but usually only one data parameter may be displayed. (Many of the pen

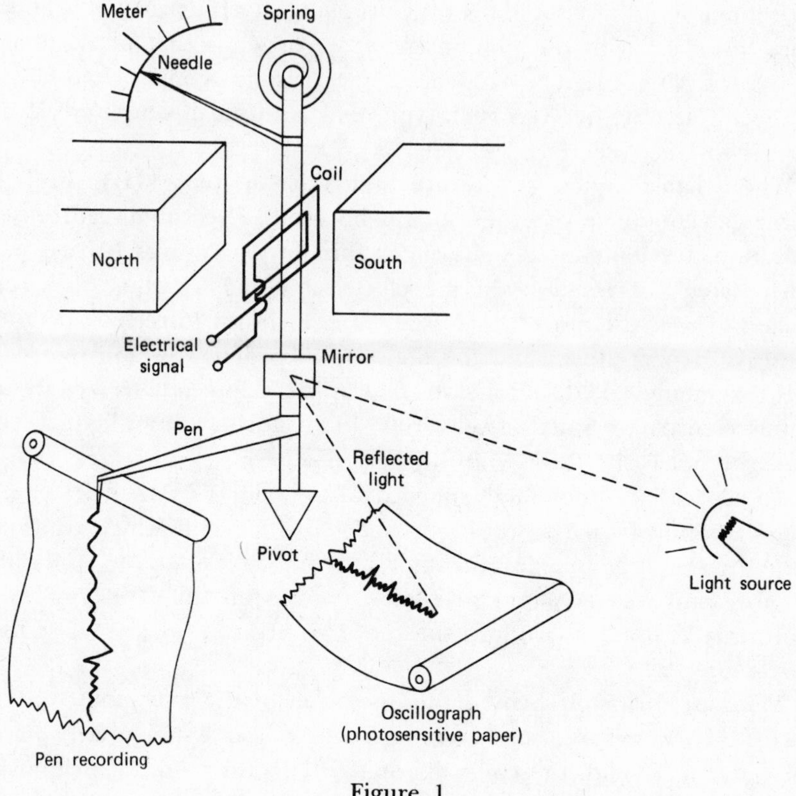

Figure 1

recorders and oscillographs used in the aerospace industry contain as many as 10 pens and 32 mirror-type galvanometers.)

It must be mentioned, though, that each galvanometer-type recorder suffers from a common problem—the normal movement of their points or light spots is not in a straight line but along a curve. With potentiometric strip-chart recorders (similar in appearance to a pen recorder) this problem does not exist because of a position feedback control system. With the potentiometric recorders, linearity errors fall into the order of ½ to 1% or better.

In physically impressing a data trace onto a chart, there is the light beam-to-photosensitive paper (or film) technique for oscillograph recorders but a variety of options for the potentiometric or pen recorder. These options include the capillary action ink-to-paper system, the felt tip pen, the pressurized ink system, the heated stylus, and the electrostatic technique.

Capillary action pens are inexpensive but at low writing speeds or when stopped can run or drip. Felt tip pens wear out rapidly and in doing so deliver an ever-broadening ink trace to a recorder's chart paper. Although not so susceptible to clogging as the capillary action pen, the felt tip is also a "potential leaker" at low writing speeds.

Pressurized ink systems deliver just the right amount of ink per writing speed to render superb chart displays. Such systems are quite costly, however, and require special paper. The heated stylus system is one in which the "pen" is actually a heated piece of metal that burns traces onto moving wax-coated paper.

The electrostatic technique entails generating a trace with a high-voltage stylus that breaks down a light-colored coating material (usually gray) covering a black carbon-filled base material. Although neither the heated stylus nor the electrostatic record can match the pen recorder in terms of contrast, the cost of these systems is usually much lower.

Also noteworthy of mention are two other systems, the *X-Y* and the *X-Y-Y'* plotter and the liquid jet ink writing system. The plotter is actually a large-sized pen recorder that records one (*X-Y*) or two (*X-Y-Y'*) traces on an 8½ x 11 or 11 x 17 sheet of paper. Trace recording is usually done by the capillary action or felt tip pen. *X-Y* and *X-Y-Y'* plotters exhibit an extremely low frequency response (i.e., a few cycles per second); their trace range is in the order of ±4 in. or more. The "penless ink squirter" directs a fine jet of ink onto paper. Its frequency response usually falls somewhere between the pen recorder and the mirror-type galvanometer.

Sample outputs from the pressurized ink pen recorder and the mirror-type galvanometer are shown in Figures 2 and 3. In selecting an analog display recorder, one must first specify the maximum frequency of expected data, the amount of recording error allowable, and the minimum number of parameters to be displayed simultaneously on the same chart.

In the fields of remote sensing, vibration analysis, and undersea acoustic data analysis the fiber optics recorder is ideal when only one measurement is to be displayed and when that measurement contains high-frequency data components in the order of several thousand

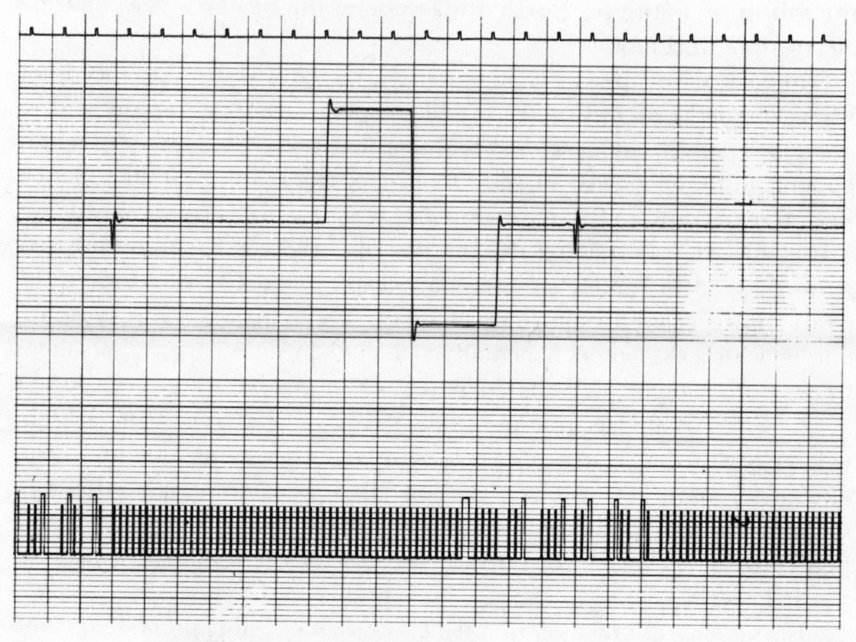

Figure 2. Sample output: pen recorder.

hertz (Figure 4). The single-channel fiber optics recorder can record both transverse and longitudinal signals. Transverse recordings are effected by applying the data signal through the cathode ray tube (CRT) vertical amplifier; the time base is represented by the horizontal sweep (Figure 5).

In longitudinal recording the data signal is applied by a horizontal amplifier. The signal is recorded in the direction of chart paper or film. Chart film speed is the time base, as it is in the conventional galvanometer oscillograph or pen recorder (Figure 6).

Among the many uses for the fiber optics recorder is in generating *spectrograms* containing amplitude, frequency, and time information on a single display (Figure 7).

In the area of undersea acoustic studies different subsurface vehicles and marine species emit unique signals, that is, spectral density characteristics. Thus spectrograms, generated from data acquired by an undersea hydrophone, can be quite useful in monitoring undersea events and can in fact be used to establish the time of occurrence of each unique signature. Spectrograms, for example, can show the pres-

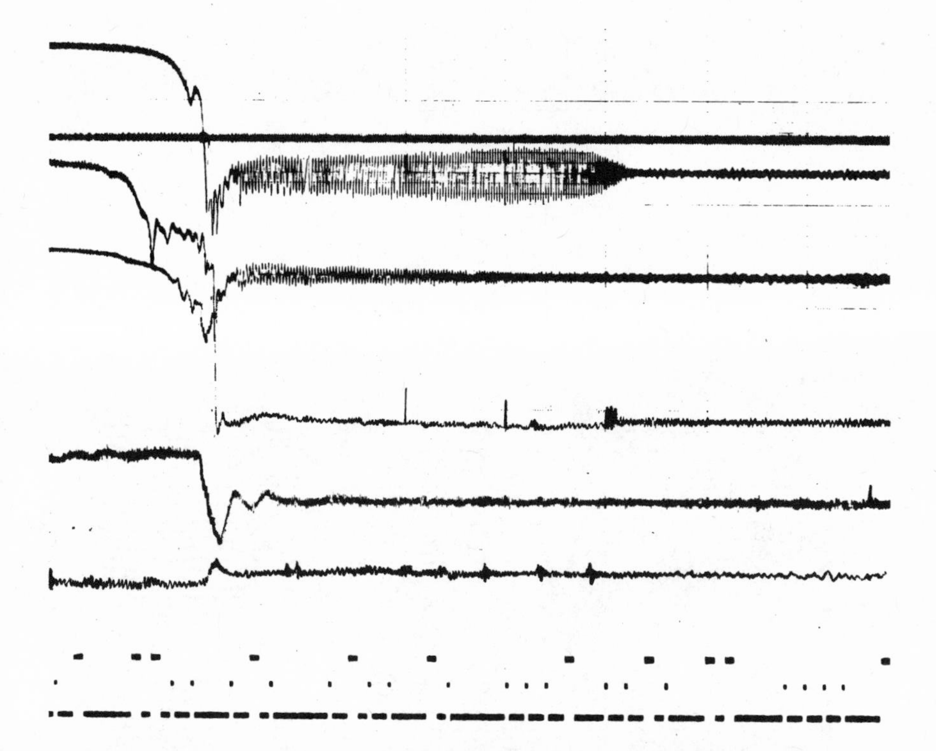

Figure 3. Sample output: oscillograph.

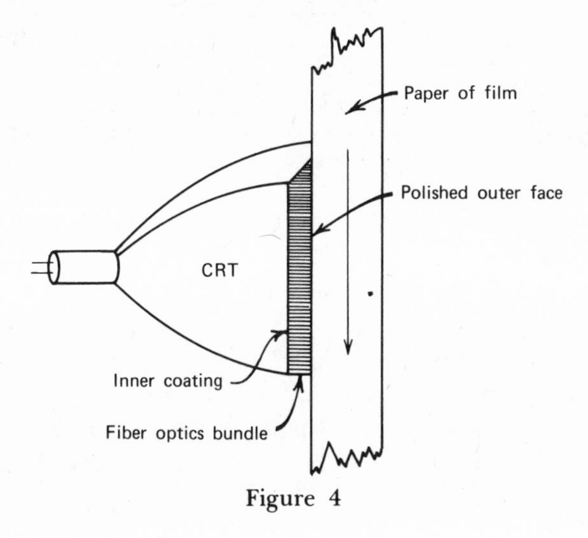

Figure 4

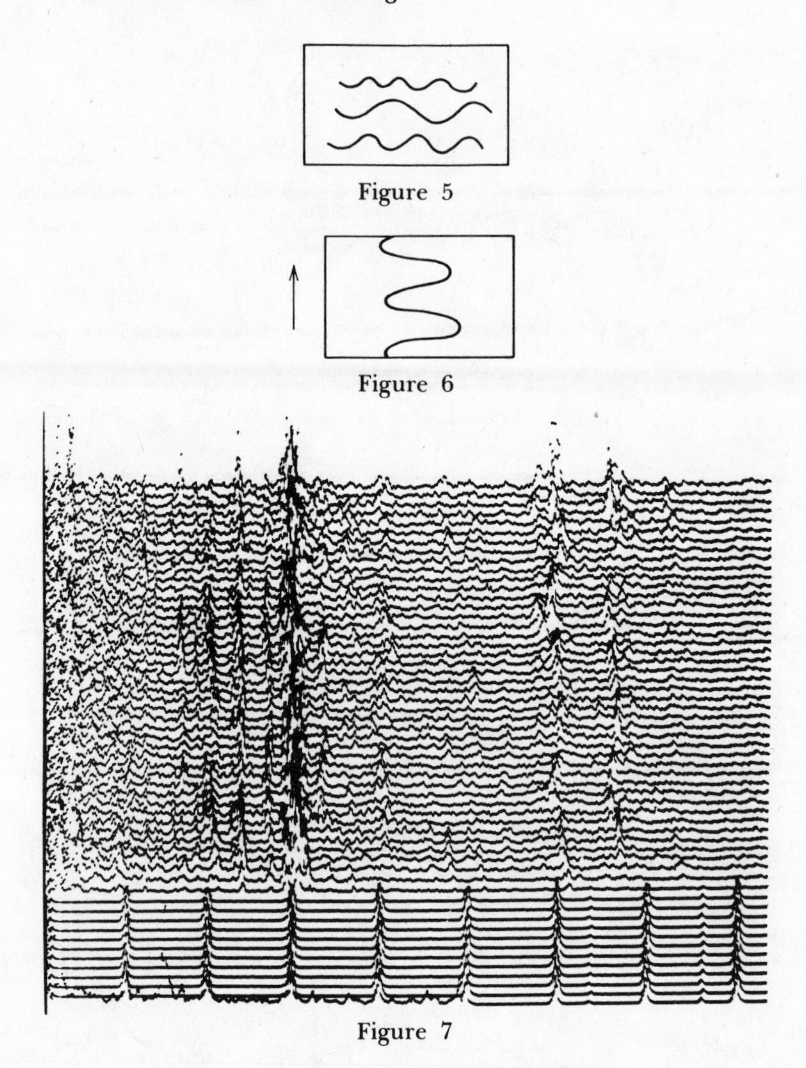

Figure 5

Figure 6

Figure 7

ence of a submarine or a school of fish within the reception range of a hydrophone.

The fiber optics recorder also serves as a means of recording earth-resources data acquired by air- or spacecraft multispectral or thermal infrared scanner. Scan-line data, similar to that presented to a television screen, may be routed directly to the fiber optics recorder. Just as in television, contrast changes may be made to enhance the CRT image that will ultimately be recorded on photosensitive paper or film.

9

DIGITAL PULSE CODE MODULATION (PCM) DATA

When large quantities of data are to be acquired by transducers and/or sensors, it is often desirable from a cost and technical standpoint to convert these data to a format suitable for processing by digital computer systems. PCM is such a format.

Satellite multispectral scanner data, for example, are generated by first sampling video outputs from each (color) detector and then generating a pulse amplitude (PAM) waveform. PAM data are routed to an analog-to-digital converter for signal encoding into a PCM waveform. Similarly, during large-scale ground and aircraft transducer data acquisition programs analog transducer output signals may be subjected to the PAM and encoding process.

In order to understand the pulse code modulation recording technique fully one must first master the basics of pulse amplitude modulation and digital-to-analog and analog-to-digital conversion. Discussions of these subjects are contained in this chapter.

PCM data reduction requires digital hardware. Although it is not the purpose of this book to discuss in depth either computer programming or computer hardware, computer number systems are covered in Chapter 10. Note that an important prerequisite to the mastering of digital recording and processing techniques is a knowledge of binary arithmetic; see Chapter 10 for binary fundamentals. Also covered are the octal and hexadecimal systems, each extensively utilized today by programmers and analysts working with PCM or computer tape data.

The primary difference between analog and digital data is that digital data are actually samples of transducer or remote sensor outputs. Analog signals, on the other hand, are continuous. No sampling is involved. When sampling occurs, it is the result of a process, defined later, called commutation.

PULSE AMPLITUDE MODULATION (PAM)

Frequently the repetitive sampling of a transducer or sensor output signal at a sampling rate that equals two or more times the highest expected data frequency will permit the accurate reconstruction of the original waveform at some later time. The first two examples in Figure 1 demonstrate this point. This third example shows that a low sampling rate is inadequate. The higher the sampling rate, the closer

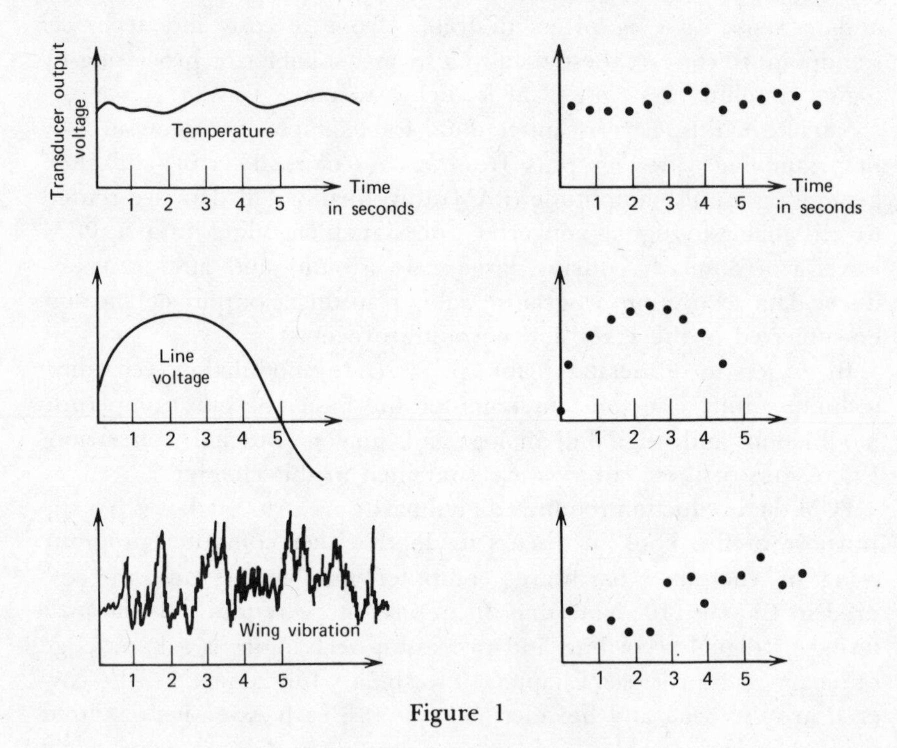

Figure 1

the reconstructed data to the original. Slowly varying signals need not be sampled so often as fast-changing ones.

Data sampling, achieved by an electronic or electromechanical commutator, permits the generation of one analog signal containing many individual samples from different transducers or sensors. That signal may then be analog-to-digitally converted into a pulse code modulated waveform or routed to a VCO for creation of PAM/FM suitable for telemetry or tape recording.

In the example given in Figure 2 the electromechanical commutator

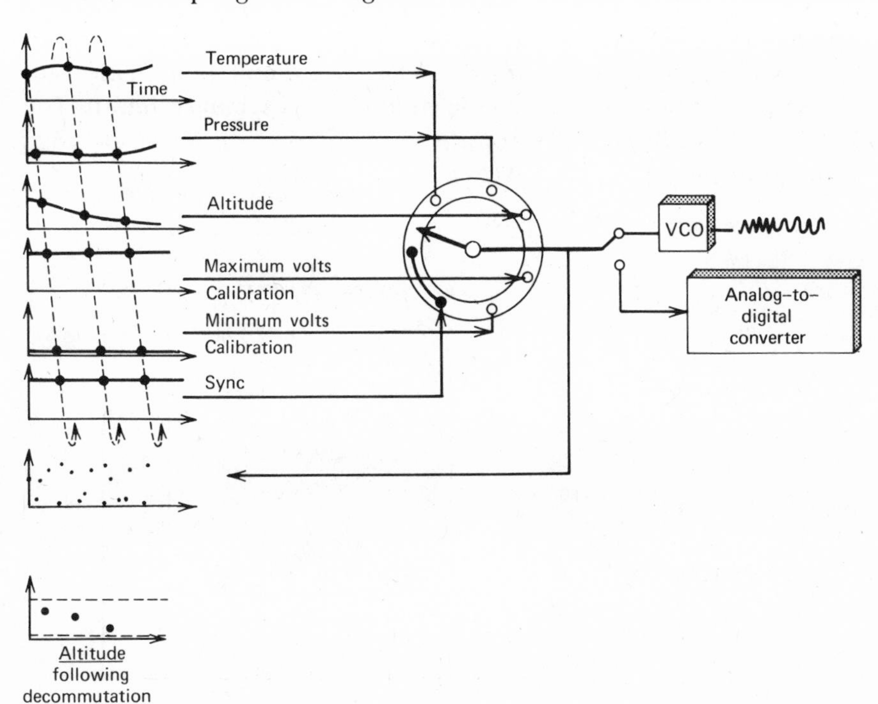

Figure 2. Important definitions associated with the PAM process include *frame*—one complete revolution of the commutator; *frame rate*—number of frames sampled per second; *sync pulse*—by referencing the various transducer inputs to their position relative to the *sync* pulse the reproduction system at a data processing station can locate the output of each specific transducer [e.g., altitude is always the third signal ("dot") to the left of the sync pulse].

is sampling three transducer signals. Note also the addition of synchronization (sync), minimum, and maximum calibrations. One commercially available PAM commutator contains room for 90 inputs. It is often advantageous from an economic standpoint to use a PAM system rather than, in this case, 90 VCOs. PAM, however, is possible only when data signals are low enough in frequency to permit accurate sampling. Digital data recording (e.g., creation of PCM) becomes merely a conversion of each successive PAM pulse into a binary coded waveform which is acceptable to the digital computer.

DIGITAL-TO-ANALOG CONVERSION (DAC)

The digital-to-analog converter is an electronic device which, by digitally operated switches (i.e., open or closed), generates an analog output signal proportional to the magnitude of its binary input. The switches, shown in Figure 3, control the amount of resistance switched into or out of the overall DAC circuit.

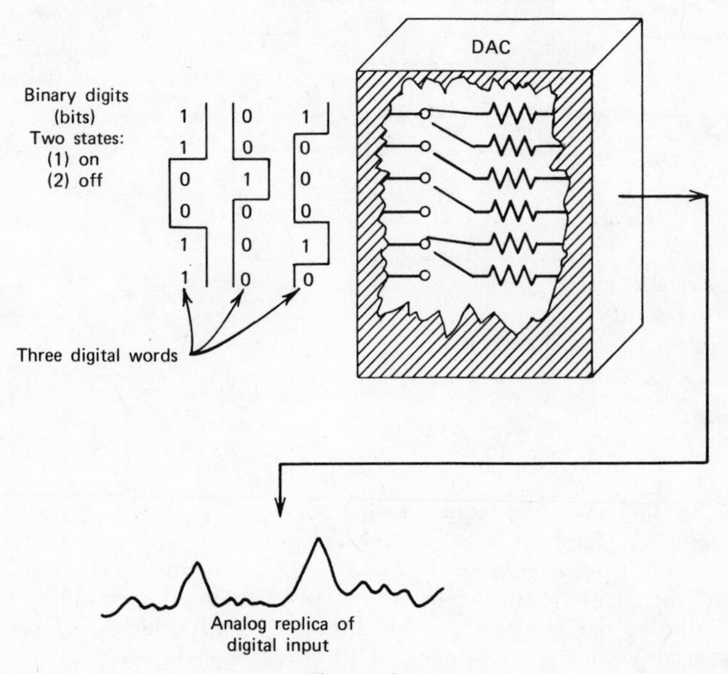

Figure 3

The DAC serves as an important part of analog-to-digital converters and the PCM system. Its basic operational principle is best described after a review of basic circuit theory.

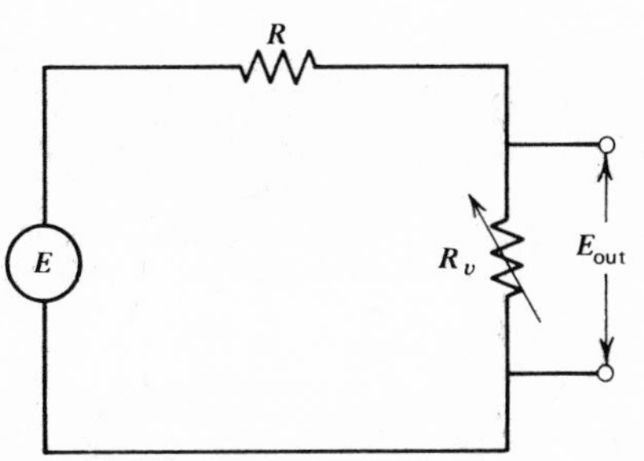

Figure 4. E = voltage source, R = resistance, R_v = variable resistor, and E_{out} = circuit output voltage.

Figure 4 is a simple series circuit that contains two resistors (one of which is variable) and a voltage source. As the variable resistance approaches minimum, the output voltage approaches minimum. Conversely, as the variable resistance approaches maximum, the output voltage approaches maximum.

Figure 5 illustrates a basic DAC. Unlike the circuit shown in Figure 4, the variable resistor is replaced by special digitally controlled resis-

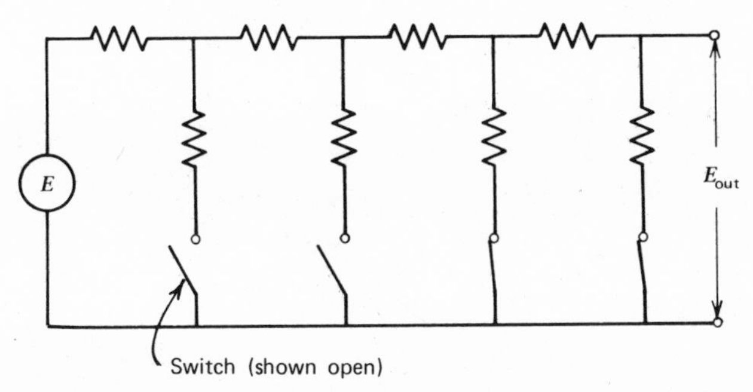

Figure 5. E_{out} is maximum when all switches are open. (In this, a parallel circuit, the more parallel resistors, the lower the output voltage.)

tors that are connected into the circuit or remain "open," depending on whether they see a 1 or a 0. The more resistors switched in, the lower the circuits' output voltage. Note that we start with a constant voltage E and that if resistors are placed in the path of flow the final voltage out will be somewhere between high and low strength. The amount and position of the 1's and 0's in each digital word determines the resistors to be connected into the circuit.

Most DACs today consist of hybrid circuits; that is, they contain both analog and digital components. One of the key analog parts of the hybrid DAC is the *operational amplifier* (Figure 6). The equation

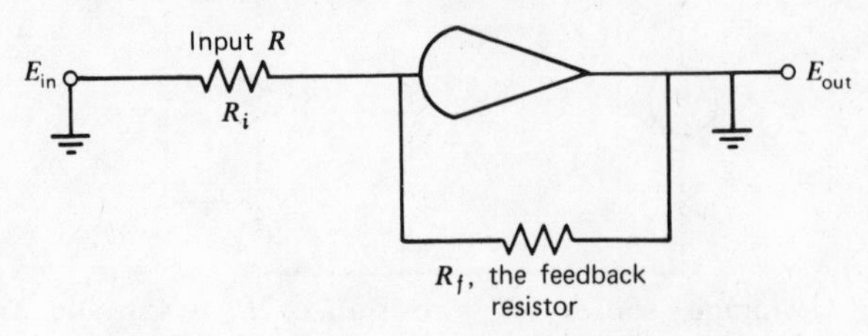

Figure 6. R_f/R_i = amplification factor.

governing the output voltage from an operational amplifier is as follows:

$$E_{out} = -E_{in}\frac{R_f}{R_i}$$

An operational amplifier with an amplification factor of 1 would produce a − 1-V output for a + 1-V input and a gain of 2 would produce a − 2-V output. Illustrated in Figure 7 is a hybrid DAC that contains the operational amplifier.

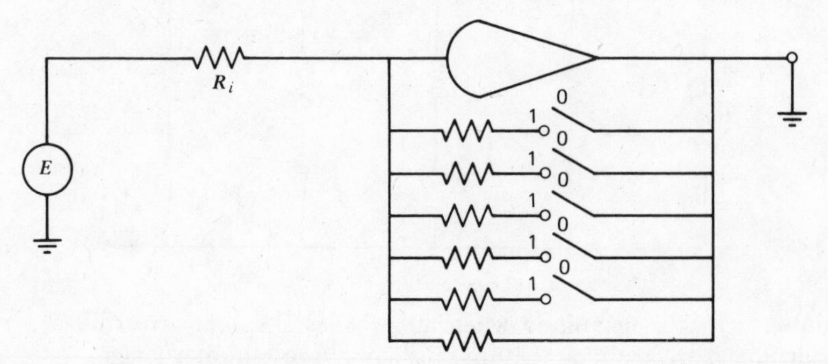

Figure 7. R_f/R_i = amplification factor.

Digital words, when connected to the on or off (1 or 0) switches will, depending on whether 1's are present, cause certain "weighting" resistors to be placed in parallel in the feedback resistor network, thereby causing specific settings of gain in the amplifier. Use of this scheme permits the analog output of the amplifier to represent the digital words connected to its input.

In actual practice, however, the mechanical switches are replaced with electronic switches and *flip flops.* These components permit high-speed switching within the DAC. The flip flop is a device that depends on the binary state stored in it and places the electronic switch to the open or closed position, depending on that state. Illustrated in Figure 8 is a 10-bit DAC capable of accepting 10-bit digital words.

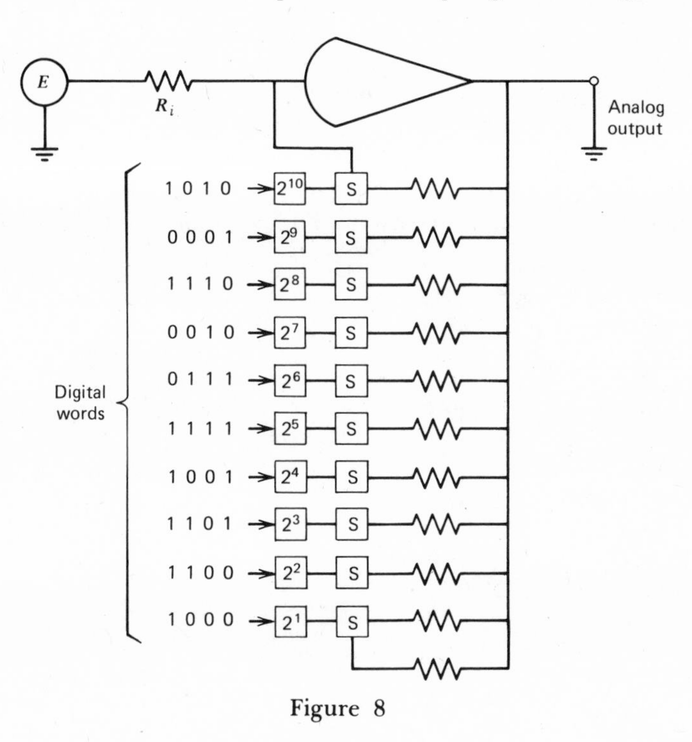

Figure 8

The analog output voltage is usually maximum when all 1's are connected to the DAC and minimum when all 0's enter.

ANALOG-TO-DIGITAL CONVERSION (A/D)

The A/D process is one in which a continuous analog signal is sampled at a rate at least twice that of the highest expected data frequency, and in which each sample is converted into a digital word (Figure 9). Theoretically, a minimum of two data samples per highest expected data frequency is sufficient. In actual practice, however, between 5 and 10 are commonly used. Excess sampling adds no appre-

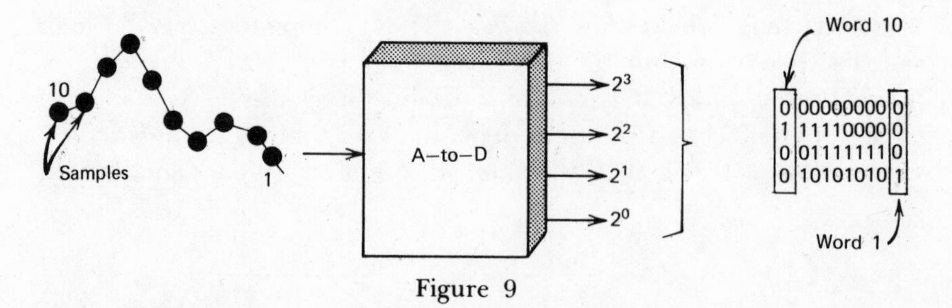

Figure 9

ciable accuracy to processed data but requires a needless excess of computer tape and time and also requires the availability of a high-frequency response tape recorder. The example in Figure 10 illus-

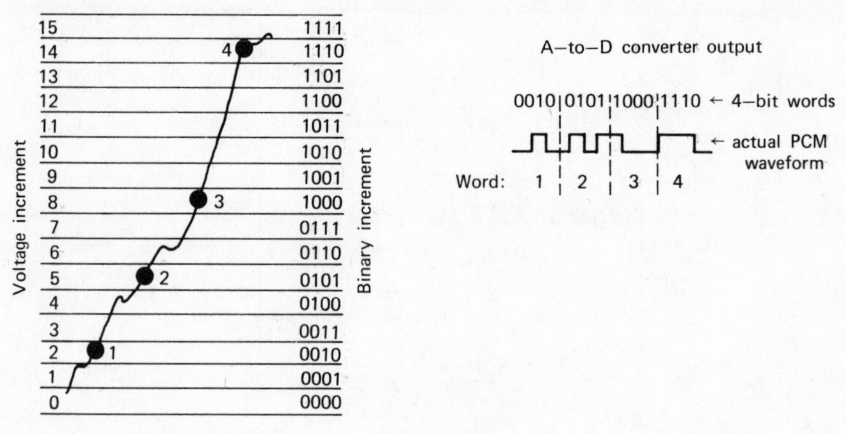

Figure 10

trates the analog sample to digital word process. The largest possible four-bit word, $1111_2 = 1 \times 2^0 + 1 \times 2^1 + 1 \times 2^2 + 1 \times 2^3 = 15_{10}$. The smallest word, $0000_2 = 0_{10}$.

Analog-to-digital converters perform a sample and hold operation; this function may be readily understood by considering a similar physical sample and hold operation. Consider first the manner in which the weight of an unknown object (sample) is determined by use of a balance scale (Figure 11). As known weights are placed on the right, the scale indicates one of three conditions: (1) less than, (2) equal to, or (3) more than. When the unknown weight on the left is equal to the proper combination of known weights on the right, the sample weight is determined and may be recorded.

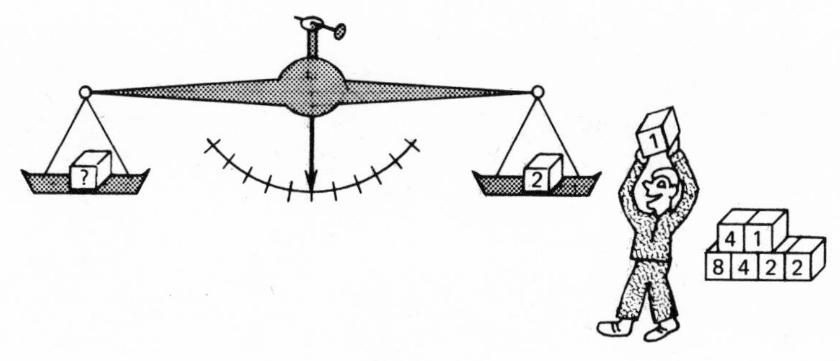

Figure 11

In the analog sample-to-digital waveform conversion process analog samples (just like the unknown weights) are placed, one at a time, into a sample and hold circuit. Meanwhile, known voltages are balanced against each sample. These known voltages are generated by a precision digital-to-analog converter which continues to update (increase value) by increments of 1_2 until a voltage comparator, like the scale arrow, indicates a balanced condition.

Once a balanced condition has been achieved the digital input to the DAC is routed to tape or computer and a new analog sample is subjected to the same process. Figure 12 illustrates conversion of an analog voltage to a four-bit word.

The time required for an analog-to-digital conversion depends on the number of decisions that must be made by the digital decision unit and the amount of time required by the sample and hold circuit to store samples.

Present state of the art allows the entire conversion process to be completed, per sample, in a few millionths of a second. The rate at which each analog signal is sampled is of extreme importance. Sampling that is too slow causes error; sampling that is too fast is costly both from a commutator cost standpoint and from the amount of computer time needed in which high sampling rates are used. The effect of high versus low sampling rates is illustrated in Figure 13. Obviously the higher the sampling rate, the more accurate the commutator's output. The basic law in the choice of sampling rates is that the commutator (at the very minimum) sample at least twice the rate of the highest expected data frequency.

It is normally required that more than one transducer or sensor output be processed by the analog-to-digital converter; often more

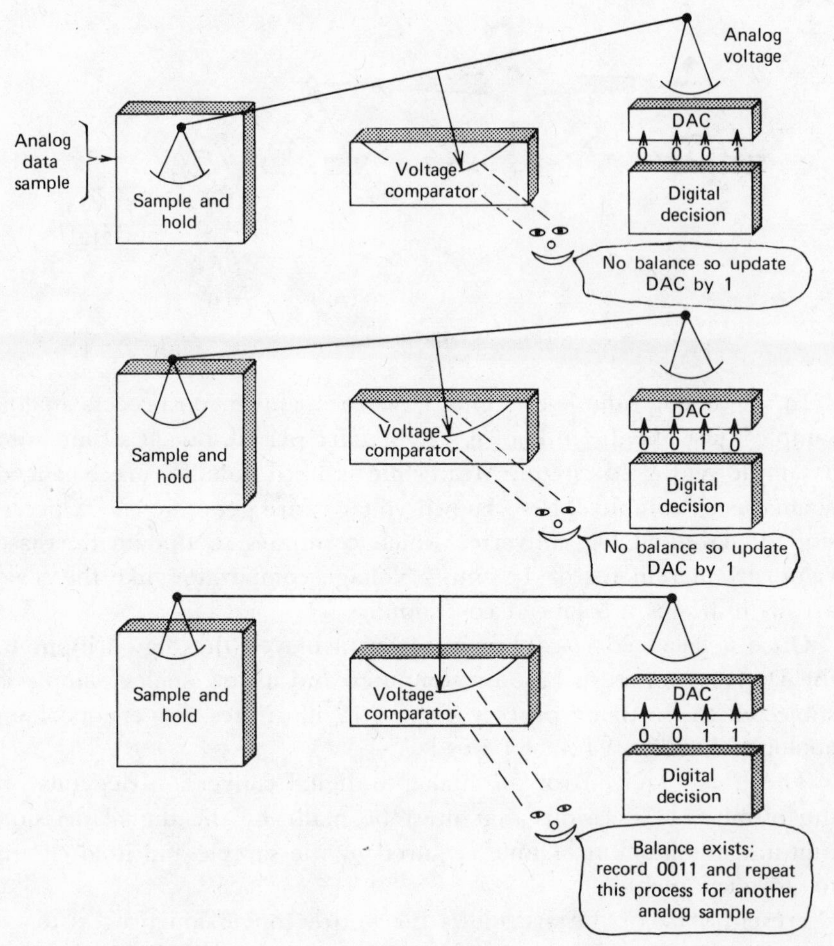

Figure 12. Conversion of an analog sample to a four-bit digital word.

than 50 signals require processing. The converter, however, cannot act on 50, 40, or even two samples at once; one signal at a time is the limit.

In order to achieve conversion of several signals, the analog-to-digital conversion system must first accept all inputs and route them, one sample at a time, to actual conversion hardware. This process requires the sampling of each signal at a discrete time and the presentation of the chain of samples to the converter. This process, commonly referred to as *time division multiplexing,* causes the creation of a

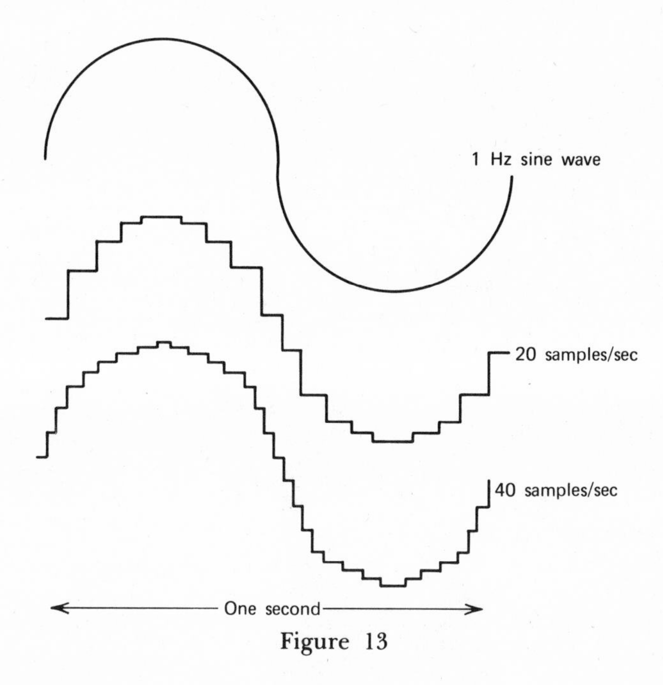

Figure 13

waveform in which, for each specific time period, a specific sensor or transducer signal exists.

Figure 14 illustrates how two inputs (*A* and *B*) are sampled and also the A/D converter output that results. The first sample, at time = t_1, is taken from waveform *A*. The second sample, at time = t_2, is taken from waveform *B*. A/D outputs occur in the same order; that is, alternating samples of *A* and *B*.

In actual practice the commutation process is performed by a device known as a multiplexor or MUX. This device consists of a series of controlled switches, each connected to an analog transducer or sensor output or a conditioned output. The logic controlling the switch selection permits only one switch to be closed at a time; all others are open. Switch control is commonly performed by the digital computer. An example that describes conversion of five analog waveforms is given in Figure 15. Sensor or transducer outputs are connected by five separate channels to the multiplexor. By computer-control analog signal 1 is processed first. A coded signal then arrives from the computer at the logic control section; here only switch 2 is closed to permit analog-to-digital conversion of signal 2. This process continues, but no more than one switch is closed at any time.

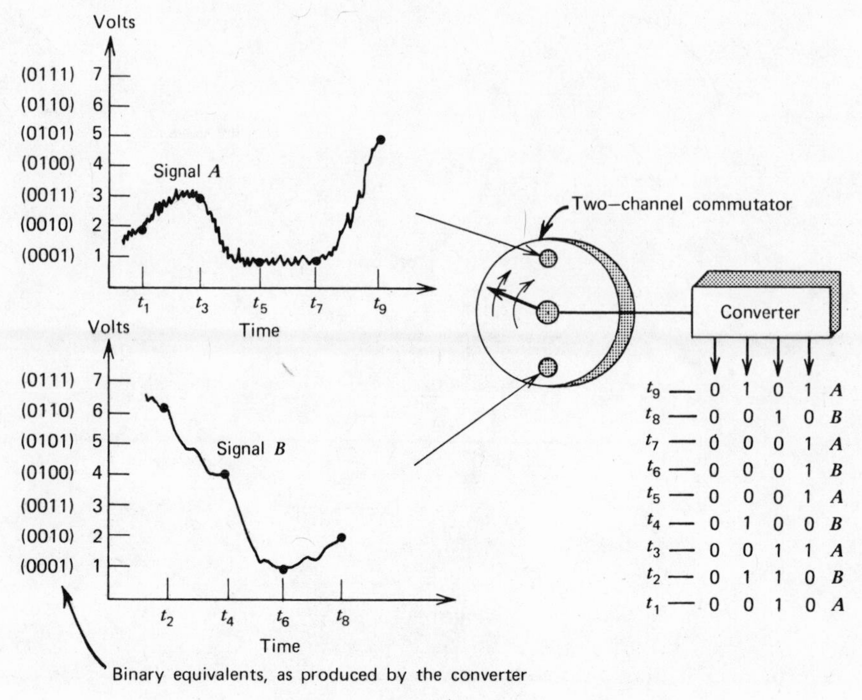

Figure 14

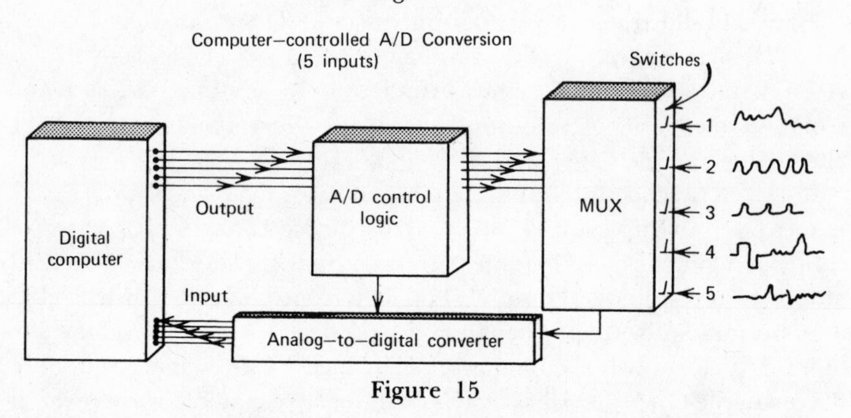

Figure 15

Out of the A/D converter come groups of binary digits (bits), each representing in digital format the voltage value of the original analog sample. Bit groups are commonly referred to as digital *words*.

Pulse code modulation (PCM) systems, discussed in the section that follows, contain analog-to-digital converters. PCM data, whether taped or by telemetry, have undergone analog-to-digital conversion at the test site—on the ground, in the air, in space, or on a planetary body.

PCM DATA SYSTEMS

Use of the pulse code modulation technique is often the only effective means of recording and transmitting huge quantities of transducer and sensor data. Unlike the FM recording, the PCM waveform contains data samples—not continuous individual signals.

PCM pulses, groups of which are termed *data words,* are binary coded; each pulse represents a 1 or a 0 (Figure 16). PCM data acquisi-

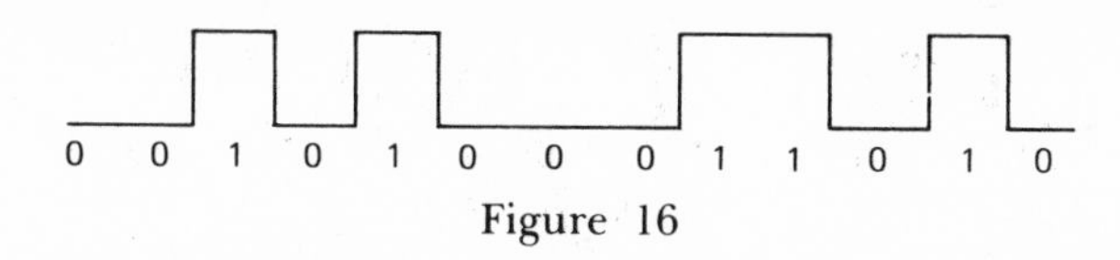

Figure 16

tion is perhaps most easily explained in a discussion of an "old fashioned" mechanical commutator. (Modern commutators are solid state and sample electronically). Routed to the pins of a commutator are analog signals. The commutator output (i.e., a pulse amplitude modulated waveform) is then routed to an analog-to-digital converter. Here each PAM pulse is converted to a digital data word which is routed, as shown in Figure 17, to a tape recorder, to a VCO for creation of PCM/FM suitable for telemetering, or directly to a computer. PCM signals are received at the data processing facility in the form of binary coded pulse trains. Each train contains time periods and each period is coded as a 1 or a 0 bit (binary digit). Each PCM data word is defined as a specified number of bit periods and each period contains a 1 or a 0 bit; each bit represents a power of the number 2.

$$\text{Bit period} = n, \ldots, 6, 5, 4, 3, 2, 1$$
$$\text{Representative value} = 2^{(n-1)}, \ldots, 2^5, 2^4, 2^3, 2^2, 2^1, 2^0$$
$$(\text{Numerical value})_{10} = \ldots 32, 16, 8, 4, 2, 1$$

n is the word length in bits. The 2^0 position (bit period 1) is the *least significant bit* (LSB) and the $2^{(n-1)}$ position contains the *most significant bit* (MSB). In most cases the least significant bit appears to the right.

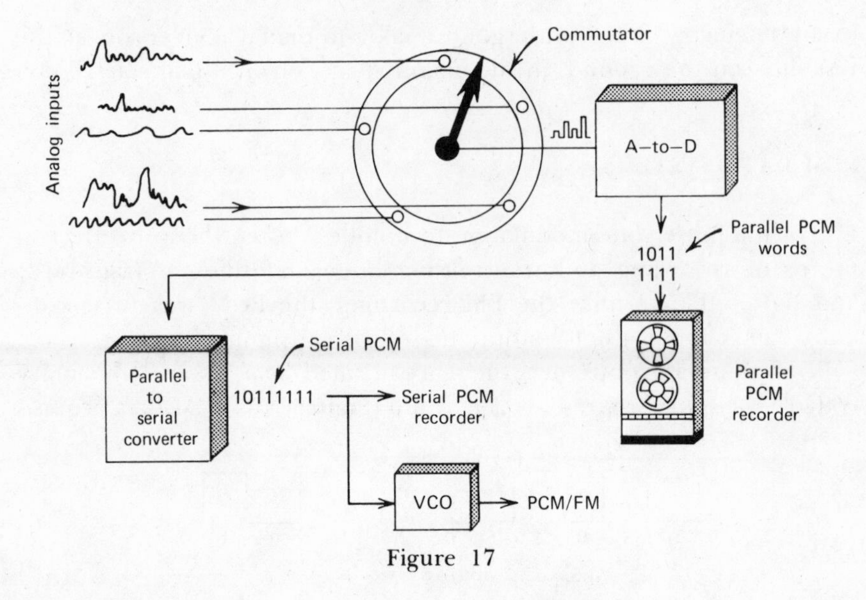

Figure 17

Example. The word value for the waveform in Figure 18 is $1 \times 2^6 + 1 \times 2^4 + 1 \times 2^3 + 1 \times 2^1 = 90$. This waveform is the nonreturn to zero (NRZ) type because the signal never returns to zero unless a zero occurs. Advantages in the use of a PCM system include the capability for recording many data signals (often hundreds), elimination of the need to perform analog-to-digital conversion at the data processing site, and a low chance of noise distorting data because decommutation hardware looks only for the presence or absence of a pulse. Problems associated with PCM use include hardware complexity, processing costs, and synchronization pulse recognition.

Supercommutation

As mentioned earlier, the basic rule in the choice of sampling rates is that (at the very minimum) the rate be at least twice that of the

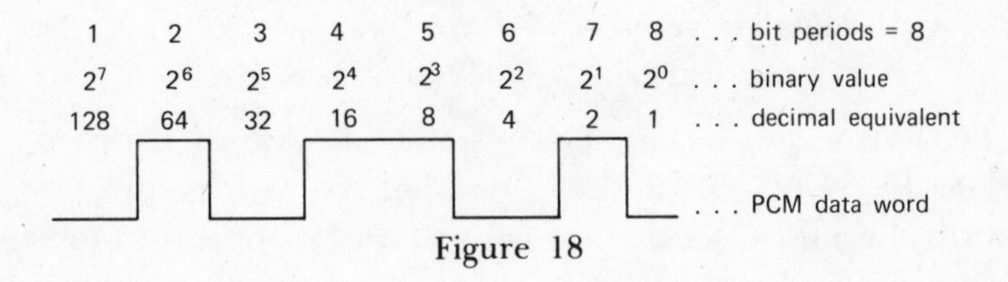

Figure 18

highest expected data frequency. Also mentioned with regard to the number of samples per highest expected data frequency was that in actual practice between 5 and 10 are used.

It would seem, then, that when a data signal's frequency is faster than that of the commutator the signal cannot be accurately recorded. This assumption is correct. However, because of a special "trick" called *supercommutation,* we may place the high-frequency signal accurately into the final PCM waveform.

Example. Measurements to be recorded and highest expected measurement data frequency are the following:

1. Temperature approx 1 Hz max
2. SO_2 level approx 1 Hz max
3. Water flow rate approx 2 Hz max

Assume that we have a commutator with room for four data inputs and a commutator frame rate of 8 frames/sec. (Each input pin is sensed 8 times/sec.) Assume also that we require a minimum sampling rate of 5.

By connecting temperature and the SO_2 level each to one pin, each of these measurements would be sampled 8 times/sec. Because the highest expected data frequency for each is about 1 Hz, accurate sampling may then be achieved; accurate sampling is five or more times the highest expected data frequency.

What happens if flow is also connected to one pin? It, too, would be sampled 8 times/sec but this is not sufficient because accurate sampling must be five or more times 2, the highest expected data frequency. Therefore connecting flow to one pin is unacceptable.

Now let us see what would happen if flow were connected to two pins (Figure 19). Here flow is sampled twice per frame and, because the frame rate is 8, 16 samples/sec would occur. Sixteen is adequate because it is greater than the minimum requirement of 10 samples/sec. By this method flow can be sampled at a rate faster than that of the commutator's frame rate. This process, in which one measurement is connected to two (or more) pins on a commutator, is known as *supercommutation.* It permits us to sample high-frequency signals often enough not to violate sampling accuracy rules.

When the number of pins on a commutator is greater than the number of samples to be recorded, the extra pins may be used to

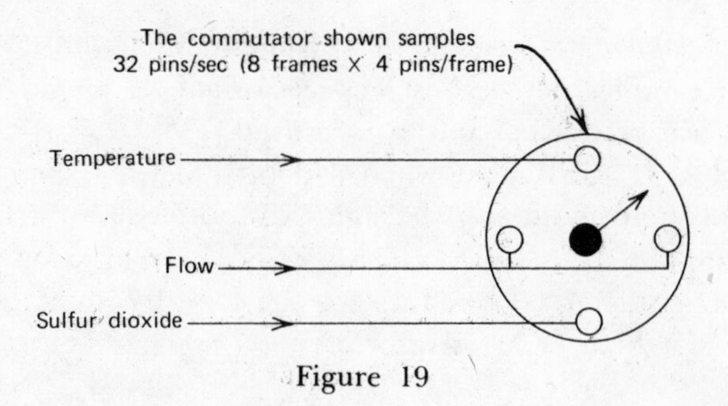

Figure 19

establish supercommutation, the means for increasing sampling accuracy for highest frequency data signals.

Subcommutation

Consider the case in which X sensor or transducer outputs are to be digitally recorded but in which only Y commutator pins are available (X is greater than Y). A shortage of pins exist.

Example. Measurements to be sampled and the highest expected measurement change rate are temperature (1/5 Hz), salinity (1/10 Hz), and relative humidity (1/20 Hz). Assume that the only available commutators contain two data pins each.

By subcommutation, that is, by the utilization of more than one commutator (a primary and one or more subcommutators connected to the primary), extra data can be added to the primary commutator output (Figure 20).

If the frame rate of the primary commutator is 1 rev/sec (i.e., 2 pins) and the frame rate of the subcommutator is ½ rev/sec, note what occurs during a 4-sec study of the main (primary) commutator output (Figure 21).

Assuming that we are using subcommutation, that is, a second commutator whose output is connected to a main commutator pin, what is our final sampling accuracy?

If accurate sampling requires sampling at five (or more) times the highest expected data frequency, then

> for temperature, accuracy requires five (or more) times 1/5 Hz (min of 1 sample/sec),

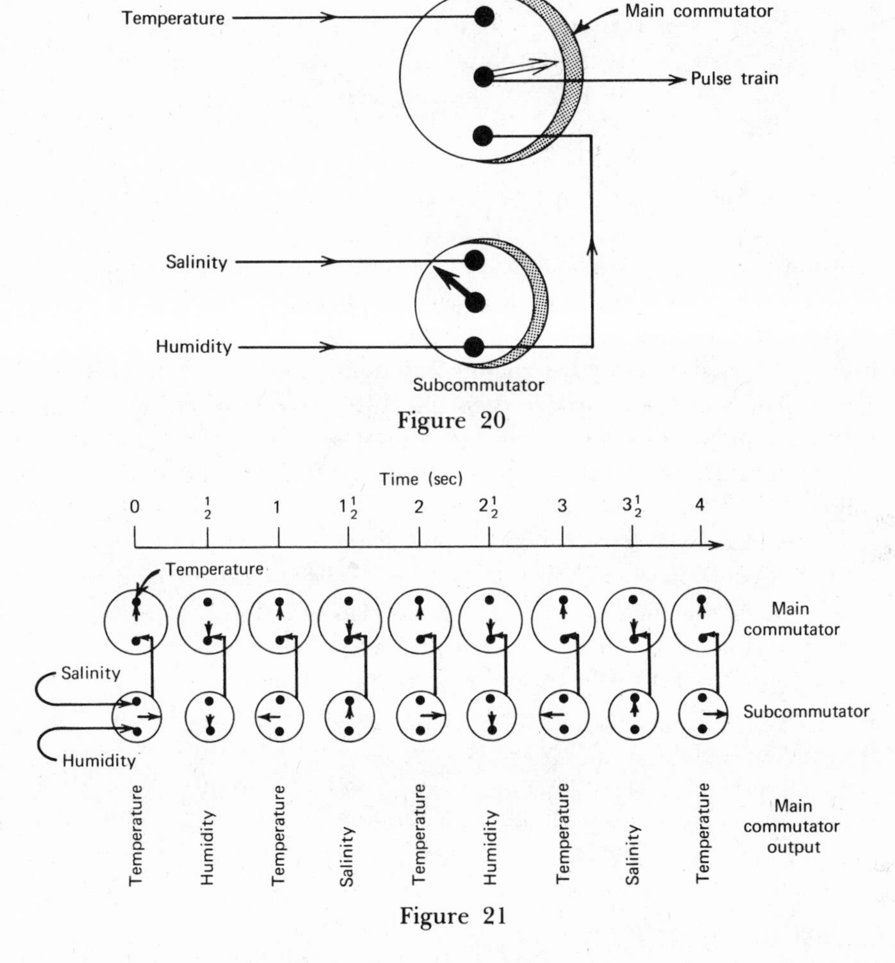

Figure 20

Figure 21

for salinity, accuracy requires five (or more) times 1/10 Hz (min of 1 sample/2 sec), and
for humidity, accuracy requires five (or more) times 1/20 Hz (min of 1 sample/4 sec).

As we can see, temperature is sampled once a second, salinity once every two seconds, and humidity also once every two seconds. Each measurement is indeed accurately sampled.

Use of subcommutation allows us to conserve pins on the main commutator and by employing a subcommutator to place "extra" measurement information into the main commutator output pulse train. Without subcommutation we should be forced to create two pulse trains from two commutators and to use two separate PCM recording systems. With subcommutation we still need two commutators but only one PCM pulse train recording system.

The Sync Pattern and the Frame Counter

Assume that we have an eight-pin commutator with a two-pin subcommutator affixed to the sixth pin on the main commutator (see Figure 22). As mentioned earlier in this chapter, whenever a commutator is used, pins on the main commutator must be reserved for a sync word. This sync word, in the final PCM waveform, will serve as a reference point from which all data words may be found. (Example: air temperature might always be the fifth word after sync and salinity might always be the sixth.)

Let us place sync information on pins 1 and 2. This information (i.e., a specific constant voltage input) will eventually be transformed into a binary word that is different from each data word and repeats itself once per frame. Now let us connect frame count detection circuitry to pin 3. This circuitry will cause the creation of a binary word that will tell us how many frames have been completed and is used to identify the subcommutator channel being recorded during each frame. Four data signals are placed on pins 4, 5, 7, and 8 and two more data signals on the two pins of the subcommutator. Figure 22 illustrates each of these connections and, more precisely, a typical pulse code modulation (PCM) data acquisition system.

Final Output

Digital words created by the analog-to-digital conversion hardware (Figure 23).

COMMON PCM FORMATS

Three commonly utilized PCM formats are illustrated in Figure 24. When low-frequency response recorders are available, the NRZ code is appropriate. When noise problems cannot be avoided in recording, the split phase code is often recommended. RZ is generally not in

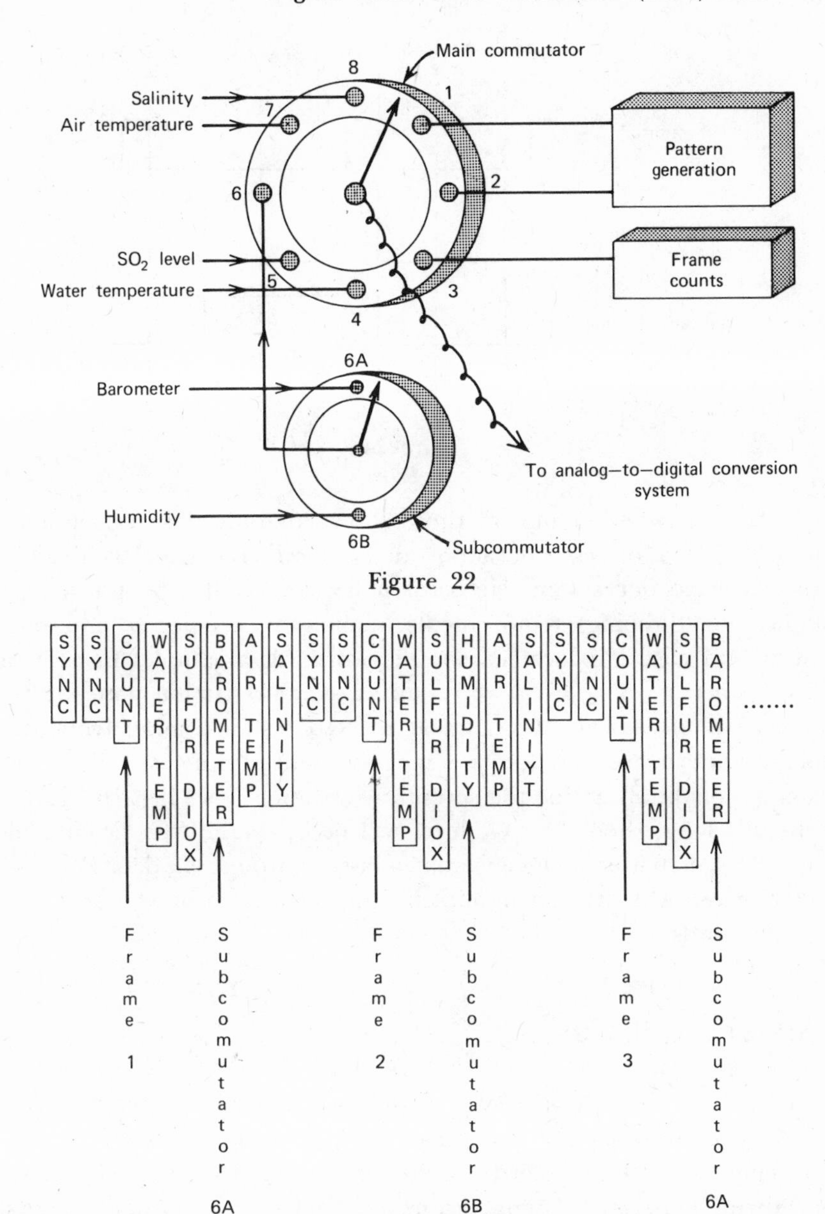

Figure 22

Figure 23

common use because decommutation of the RZ waveform introduces distortion and the transmission of a series of "zeros" does not easily tell us where to find the start and end of a bit period. Selection of a

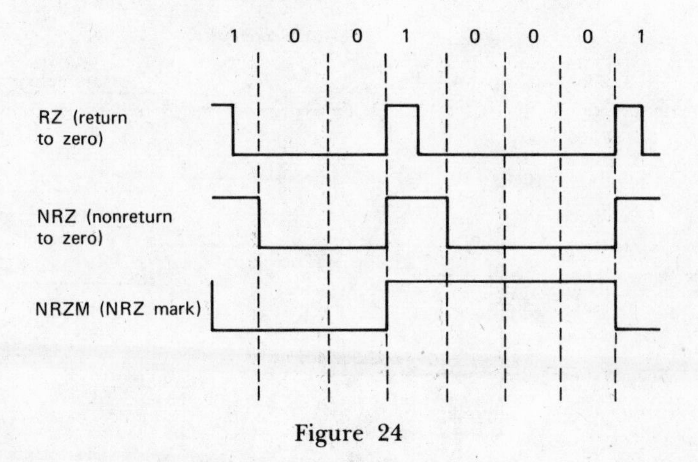

Figure 24

PCM format, whether one of those shown or one of several others, must be based on considerations that include recorder bandwidth, expected recorder system electrical noise, expected tape jitter, tape duplication (dubbing) requirements, maximum bits per second rate, and system cost. To locate a specific word (data sample) in a PCM wave train (e.g., SO_2 level) we require a synchronization pattern. Because each measurement occurs at a fixed distance away from this unique pattern, we may locate a measurement sample. SO_2 level, for example, might always be the seventh word after sync. If each word is eight bits long, then the SO_2 level will occur at the fifty-seventh bit after the sync pattern. Once we have passed through all data words in a PCM waveform, sync again appears and is followed by another *frame* of data words.

PCM DATA PROCESSING

The test data or environmental data processing facility generally receives PCM data in serial or parallel form, both shown later.

During the tape-record and tape-reproduce process serial waveforms experience some distortion; for example, when a signal should look like:

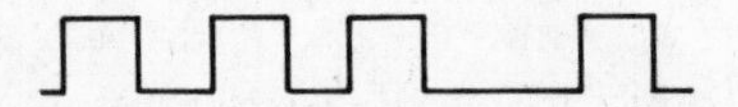

it turns out to look like:

Without conditioning, degraded serial PCM waveforms are difficult to process because the exact start and end times of pulses (binary digits) are difficult to locate. The device that conditions these inputs is called a *bit synchronizer.* Its output consists of "restored" data and timing reference markers, or *clock pulses.*

Parallel PCM, on the other hand, are easier to work with because we receive one separate word at a time, not a chain of pulses that must be separated into words. Parallel PCM, however, must be converted to a serial format in order that the sync pattern and words destined to enter a computer may be easily recognized. The device that performs this conversion is known as a *parallel-to-serial converter.* Its output consists of a serial train of pulses and accompanying synchronous clock pulses.

Serial data and accompanying clock data, whether from converted parallel or conditioned serial data, are then routed to an *input shift register,* the contents of which are compared with the contents of another, called the pattern register. Within the digital pattern register is the sync pattern. When the contents of both registers are equal, sync has been detected; data words are then expected to follow. Once sync is achieved, a *bits-per-word counter* then begins counting clock pulses. As soon as the counter reaches a predetermined number (i.e., the number of bits in a data word), the contents of the input shift register are routed to the PCM decommutation system's *format register.* The clock resets to zero and begins a count on the bits in the next data word. As this process continues, the contents of the format register (serial data words) are transferred in parallel form to the digital computer. (Computers accept parallel words only.)

Because the total number of words (and bits) in a PCM frame are known, this decommutation process can be made to deliver error messages if, between sync pulses, the total number of bits counted during a frame by the bits-per-word counter does not equal the expected total. Also, if, in the decommutation process, the pattern register and the input shift register's contents are not equal during the sync period, a loss of sync error message is generated.

——→ 1001010001101010101010101010101011110101011001010 ——→

Serial

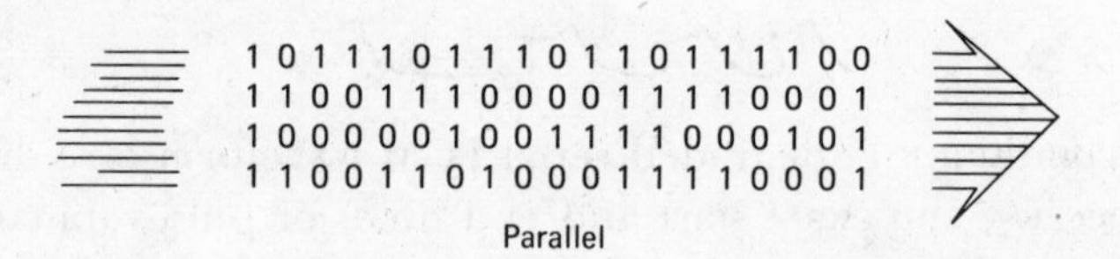

SUMMARY

In summary, the PCM data acquisition system creates a digital waveform from analog data samples. This signal, which constitutes binary "ones" and "zeros," contains

- data samples
- sync information
- frame counts

The PCM data reduction system in the electronic data processing facility conditions serial data or converts parallel data to a serial format and presents this information, with clock pulses, to an input shift register. When the contents of this register match that of a preset pattern register, sync and the start of another frame of data signals are recognized. A bits-per-word counter is then used to deliver words to a format register, from which the digital computer may extract parallel words in a format acceptable for final computer processing.

Pulse code modulation has been established as an accepted and proven means of recording the vast quantities of data that typically are found in programs of aircraft flight test, spacecraft earth-resources survey, and a host of ground test and survey programs.

10

COMPUTER NUMBER SYSTEMS

Digital data acquisition systems, digital tape recorders, and digital data processing systems, including the computer, operate in what is called the binary mode. Although transducer or remote sensor outputs are originally analog signals, they are binary-coded if digital recording and processing is wanted.

In order to understand digital recording and processing techniques fully the fundamentals of the binary number system should be mastered. Also, because many computer programs operate in the octal or hexidecimal mode, it is recommended that the serious student of computing methods familiarize himself with these systems.

The number system we are all acquainted with is the Arabic or decimal. In the paragraphs that follow, it, as well as the binary, octal, and hexadecimal system, is discussed. Examples of arithmetic in each system are given.

THE DECIMAL NUMBER SYSTEM

- It begins with zero.
- It has 10 different symbols or a *radix** of 10.
- It is *positional*; that is, the position of a digit in the total number tells its value.

Example.

□ □ □•□ □ □ = a number with a decimal point
C B A a b c = position symbols

The value of a number = $(\text{radix})^0$ times the digit in position A +
$(\text{radix})^1$ times the digit in position B +
$(\text{radix})^2$ times the digit in position C +
$(\text{radix})^{-1}$ times the digit in position a +
$(\text{radix})^{-2}$ times the digit in position b +
$(\text{radix})^{-3}$ times the digit in position c

If the above number were 0 3 2 • 3 0 0, its value would be expressed as, $(10)^0 \times 2 + (10)^1 \times 3 + (10)^2 \times 0 + (10)^{-1} \times 3 + (10)^{-2} \times 0 + (10)^{-3} \times 0 = 2 + 30 + 0 + 0.3 + 0 + 0 = 32.3$.

Another characteristic of the decimal number system is that it is *ordered*. Given any one of the 10 symbols, we know which symbol comes before it and which follows it (e.g., 6 follows 7, 5 precedes 6).

Summarizing the four decimal-system characteristics, we have found that the system begins with zero, it has a radix, it is positional, and it is ordered. These are the characteristics of the binary, the octal, and the hexidecimal systems as well. One key rule to remember is that no matter what number system is used the radix is always one unit larger than the largest character, for example, in the decimal system the characters are 0,1,2,3,4,5,6,7,8, and 9. The radix 10 is one unit larger than 9, the largest character. In the octal system characters are 0,1,2,3,4,5,6, and 7; the radix is 8. The binary system has a radix of 2 and its characters are 0 and 1.

Examples of impossible numbers are 87_8 (base 8 indicates octal number), 547182_8, 66_2 (in the binary system), and 100201001_2. Legal numbers are 64371_8 and 1_2. Expression of decimal equivalents of numbers with a radix other than 10 can be achieved via the *polynomial expansion method.*

Example. $123_8 = \text{what}_{10}?$

$$123_8 = 3 \times 8^0 + 2 \times 8^1 + 1 \times 8^2 = 3 + 16 + 64 = 83_{10}.$$

Example. $10.12_{16} = \text{what}_{10}$ (base 16 indicates hexadecimal system)?

$$10.12_{16} = 0 \times 16^0 + 1 \times 16^1 + 1 \times 16^{-1} + 2 \times 16^{-2}$$
$$= 0 + 16 + 1/16 + 2/256$$
$$= 0 + \frac{4096}{16} + \frac{16}{256} + \frac{2}{256} = \text{approximately } 16.07_{10}.$$

Why are 2, 8, and 16 common systems? Part of the answer is related to basic electronics. The transistor, a component in use in many circuits, can easily operate in two states, on and off. Transistors are contained in switching circuits; on and off can be renamed "one" and "zero."

Systems with bases of 3, 5, or 1 do not possess a number of discrete values (namely two) that are relatable to the "conduct"/"no conduct" or "one"/"zero" circuit components contained digital circuits. Because of the nature of electrical components, computers operate efficiently in the binary mode, but, when a computer programmer becomes involved with arithmetic operations, he prefers to use an octal or hexadecimal system because the higher the radix, the fewer the programming steps required.

Hardware (digital data aquisition and processing systems) therefore operates in binary circuits, but programming and the reading of information into and out of a digital system is frequently done in the octal or hexadecimal mode. The chart that follows indicates decimal numbers 1 to 21 and their binary, octal, and hexidecimal equivalents.

Decimal	Binary	Octal	Hexadecimal
1_{10}	00001_2	1_8	1_{16}
2	00010	2	2
3	00011	3	3
4	00100	4	4
5	00101	5	5
6	00110	6	6
7	00111	7	7
8	01000	10	8
9	01001	11	9
10	01010	12	A
11	01011	13	B
12	01100	14	C
13	01101	15	D
14	01110	16	E
15	01111	17	F
16	10000	20	10
17	10001	21	11
18	10010	22	12
19	10011	23	13
20	10100	24	14
21	10101	25	15

Technique for Converting from Radix 2, 8, or 16 to Another Radix

First, convert characters to their binary equivalent. (If the characters are in binary, leave them alone.)
Second, from the decimal point regroup the binary characters into groups of three (if going to radix 8) or to four (if going to radix 16). When extra zeros are needed to complete a group, add them.
Third, utilize the chart and write the number in the new radix.

Example. Convert 5252_8 to radix 16:

$$5252_8 = \underset{5}{101}\ \underset{2}{010}\ \underset{5}{101}\ \underset{2}{010}$$

101 010 101 010 must be converted to groups of four, or 1010 1010 1010_{16}.

$$1010\ 1010\ 1010 = AAA_{16}$$

Example. What number in radix 16 = 20230_8? From step one we convert to binary . . . 010 000 010 011 000. From step two, we regroup as follows:

$$0010\ 0000\ 1001\ 1000$$

We have added a zero to complete the group. From step three:

$$\underset{2}{0010}\ \underset{0}{0000}\ \underset{9}{1001}\ \underset{8}{1000} = 2098_{16}.$$

Example. What number in radix 2 = ABC_{16}? From step one 1010 1011 1100 is the binary grouping. Second, by grouping into one binary number we obtain

$$101010111100 \leftarrow \text{the } \textit{answer}$$

which may also be written as

$$\underset{5}{101}\ \underset{2}{010}\ \underset{7}{111}\ \underset{4_8}{100}, \text{ or, } = 5274_8$$

Example. What number in radix 8 = 110100_2? From step one we leave 110100_2 alone. From step two regrouping into groups of three for octal yields

$$110\ 100$$

From the last step, using the chart, we obtain 64_8.

Example. BAD_{16} = what$_8$? From step one 1011 1010 1101 is the binary equivalent. Regrouping yields 101 110 101 101. From the chart we obtain 5658_8.

Conversion from Base 10 to Base 2, 8 or 16

The method entails successive division of the base 10 number by the new radix and the remainders form the new number. (Note that it applies to whole numbers only.)

Example. What number in base 2 = 91_{10}? Performing the successive divisions, we obtain

$$
\begin{aligned}
91/2 &= 45 + \textcircled{1}\\
45/2 &= 22 + \textcircled{1}\\
22/2 &= 11 + \textcircled{0}\\
11/2 &= 5 + \textcircled{1}\\
5/2 &= 2 + \textcircled{1}\\
2/2 &= 1 + \textcircled{0}\\
1/2 &= 0 + \textcircled{1}
\end{aligned}
$$

By grouping the remainders we obtain the answer: 1011011_2. The most significant character (the one to the left) is always the last remainder obtained.

Example. 12345_{10} = what 16? By performing successive divisions we have

$$
\begin{aligned}
12345/16 &= 771 + \textcircled{9}\\
771/16 &= 48 + \textcircled{3}\\
48/16 &= 3 + \textcircled{0}\\
3/16 &= 0 + \textcircled{3}
\end{aligned}
$$

SOLUTION: 3039_{16}..

In digital system design the binary system is ideal; like electrical circuit components, it possesses two states. The computer programmer, on the other hand, prefers to use octal or hexadecimal arithmetic because fewer steps and character notations are required. Note that the number 15 in decimal is 17 in octal, *F* in hexadecimal, but a lengthy 01111 in binary.

The differences in binary, octal, and hexadecimal arithmetic operations demonstrate the reasons prompting the designer working in binary but the programmer working with another "less steps required" system.

Binary Addition and Subtraction

ADDITION RULES

```
  1    1    0    0
 +1   +0   +1   +0
 --   --   --   --
 10    1    1    0
 ↑_ this "1" is called a carry.
```

SUBTRACTION RULES

```
  0    0    1    1
 -1   -0   -1   -0
 --   --   --   --
  1    0    0    1
 ↑_ here, if another column exists to the left,
    deplete it by a "1."
```

Examples (Addition)

```
(1)   1     1 1            (2)  1 1 1 1 1          (3)      1 1
        1 0 1 1                 1   1 1 1                1 0 0 1 1 0 1 0
      + 1 0 1 1                     1 1 1              + 0 0 0 0 1 1 0 0
      ---------                   1 1 0 1              -----------------
      1 0 1 1 0                 + 1 0 0 1                1 0 1 0 0 1 1 0
                                + 0 1 0 0
                                + 0 0 1 1
                                + 0 1 1 1
                                + 0 1 1 1
                                + 0 0 0 1
                              -----------
                              1 0 1 1 0 0
```

Note that "carries" are shown as small 1's

Examples (Subtraction). Here, in performing the second column subtraction (from the right), we were forced to deplete the third column by a 1. Also, in the third column operation column four loses a 1.

```
1.
      0 0
      1̸ 1̸ 0 1
    - 0 1 1 1
    = 0 1 1 0
      0
2.    1̸ 0 0 1
    - 0 0 1 1
    = 0 1 1 0

3.    0 1 1 1
    - 0 1 1 0
    = 0 0 0 1
```

Binary Multiplication

Binary multiplication is applied in the same manner as decimal multiplication.

Example

```
          1 0 1 1
        × 1 0 0 1
          1 0 1 1   ← addition is performed by binary
        0 0 0 0       arithmetic rules, using "car-
      0 0 0 0         ries" as required.
    1 0 1 1
    1 1 0 0 0 1 1
```

Binary Division

Binary division is performed in the same manner as decimal division.

Example

```
                      1 1 0 0
    1 1 0 1 | 1 0 0 1 0 0 0 0
                1 1 0 0
                -------
                  1 1 0 0
                  1 1 0 0
                  -------
                  0 0 0 0
```

Example

```
              1 0 0 1 & 1/1 0 0 1
1 0 0 1 | 1 0 1 0 0 1 0
          1 0 0 1
          -------
          0 0 0 1 0 1 0
                1 0 0 1
                -------
                      1
```

Note that in the subtraction operation

$$\begin{array}{r} 0 \\ -\ 1 \\ \hline =\ 1 \end{array}$$

The column to the left is depleted by a 1.

OCTAL ARITHMETIC

From the chart shown earlier it can be observed that in octal counting the characters 8 and 9 are missing. Octal numbers thus 1, 2, 3, 4, 5, 6, 7, 10, 11, 12, 13, 14, 15, 17, 20, 21, and so on.

When octal is used frequently, as the computer programmer does, the octal addition and multiplication tables are rapidly mastered.

+							
0	1	2	3	4	5	6	7
1	2	3	4	5	6	7	10
2	3	4	5	6	7	10	11
3	4	5	6	7	10	11	12
4	5	6	7	10	11	12	13
5	6	7	10	11	12	13	14
6	7	10	11	12	13	14	15
7	10	11	12	13	14	15	16

×							
0	1	2	3	4	5	6	7
1	1	2	3	4	5	6	7
2	2	4	6	10	12	14	16
3	3	6	11	14	17	22	25
4	4	10	14	20	24	30	34
5	5	12	17	24	31	36	43
6	6	14	22	30	36	44	52
7	7	16	25	34	43	52	61

Note, for example, that 6 + 2 (in octal) = 10. Not "ten," but "one-zero." Also it can be seen that 6 times 5 = 36, or "three-six."

Example (Octal Addition)

$$\begin{array}{r} 622 \\ +\ 371 \\ +\ 214 \\ \hline =\ \ ? \end{array}$$

Starting with the right column we note that 2 + 1 = 3 and the 3 + 4 = ⑦. The next column brings 2 + 7 = 11. Here we "carry" a 1 and "uncarried" or remaining 1, + 1 = ②. In the last column the "carried" 1 with 6 = 7. This 7 plus the 3 = 12. Here we "carry" the 1 and "hold" the 2. This 2 plus the column's 2 = ④.

```
   1 1
    6 2 2
  + 3 7 1
  + 2 1 4
  -------
= 1 4 2 7
```

Note that The Small 1's represent "carries."

Example (Octal Multiplication)

```
   201
 × 623
 -----
 +  ?
```

First, multiplying 3 × 201, we obtain 603. Then 2 × 201 yields 402. Finally, 6 × 201 is 1406.

```
     201
   × 623
  ------
    ¹603
    402
   1406
  ------
= 145423
```

Note the carry that followed addition in the third column.

Octal Subtraction and Octal Division

Subtraction is performed via the addition table; for example, 11 − 6 is found by looking across the "six" row and finding 11. The column number, namely 3, is the result.

A more complicated example, say $663_8 - 144_8$, is performed as follows:

Example (Octal Subtraction)

$$\begin{array}{r} 663 \\ -\ 144 \\ \hline =\ ? \end{array}$$

In the right column we cannot subtract 4 from 3. The middle column therefore must be depleted by 1 so that 4 can be subtracted from 13 (14 − 4 = ⑦). In the middle column 4 from the 5 that remains is ①. In the last column 6 − 1 = ⑤. The result is 517_8.

To perform octal division the multiplication table may be used: 36 divided by 6 is 5 via the charts. A more complicated example follows:

Example (Octal Division)

```
       15
13 | 222
      13  . . . the result of 13 times ①
      72 . . . the 7 derives from the subtraction of 13 from 22.
      67 . . . the result of multiplying 13 by ⑤.
       3
```

The result is $(15 \text{ and } 3/13)_8$.

HEXADECIMAL ARITHMETIC

In the first chart shown in this chapter it may be noted that the decimal numbers 10, 11, 12, 13, 14, and 15 were, in the hexadecimal system, *A, B, C, D, E,* and *F*. Also, the hexadecimal equivalents for decimal numbers 16, 17, and 18 are 10, 11, and 12.

Hexadecimal numbers, with their alpha and numeric characters, are more difficult to add, subtract, multiply, and divide than octal numbers, but, as in octal, addition and multiplication tables are useful.

Hexadecimal Addition and Multiplication Tables

+	0	1	2	3	4	5	6	7	8	9	A	B	C	D	E	F
0	0	1	2	3	4	5	6	7	8	9	A	B	C	D	E	F
1	1	2	3	4	5	6	7	8	9	A	B	C	E	D	F	10
2	2	3	4	5	6	7	8	9	A	B	C	D	E	F	10	11
3	3	4	5	6	7	8	9	A	B	C	D	E	F	10	11	12
4	4	5	6	7	8	9	A	B	C	D	E	F	10	11	12	13
5	5	6	7	8	9	A	B	C	D	E	F	10	11	12	13	14
6	6	7	8	9	A	B	C	D	E	F	10	11	12	13	14	15
7	7	8	9	A	B	C	D	E	F	10	11	12	13	14	15	16
8	8	9	A	B	C	D	E	F	10	11	12	13	14	15	16	17
9	9	A	B	C	D	E	F	10	11	12	13	14	15	16	17	18
A	A	B	C	D	E	F	10	11	12	13	14	15	16	17	18	19
B	B	C	D	E	F	10	11	12	13	14	15	16	17	18	19	1A
C	C	D	E	F	10	11	12	13	14	15	16	17	18	19	1A	1B
D	D	E	F	10	11	12	13	14	15	16	17	18	19	1A	1B	1C
E	E	F	10	11	12	13	14	15	16	17	18	19	1A	1B	1C	1D
F	F	10	11	12	13	14	15	16	17	18	19	1A	1B	1C	1D	1E

×	0	1	2	3	4	5	6	7	8	9	A	B	C	D	E	F
1	0	1	2	3	4	5	6	7	8	9	A	B	C	D	E	F
2	0	2	4	6	8	A	C	E	10	12	14	16	18	1A	1C	1E
3	0	3	6	9	C	F	12	15	18	1B	1E	21	24	27	2A	2D
4	0	4	8	C	10	14	18	1C	20	24	28	2C	30	34	38	3C
5	0	5	A	E	14	19	1E	23	28	2D	32	37	3C	41	46	4B
6	0	6	C	12	18	1E	24	2A	30	36	3C	42	48	4E	54	5A
7	0	7	E	15	1C	23	2A	31	38	3F	46	4D	54	5B	62	69
8	0	8	10	18	20	28	30	38	40	48	50	58	60	68	70	78
9	0	9	12	1B	24	2D	36	3F	48	51	5A	63	6C	75	7E	87
A	0	A	14	1E	28	32	3C	46	50	5A	64	6E	78	82	8C	96
B	0	B	16	21	2C	37	42	4D	58	53	6E	79	84	8E	9A	A5
C	0	C	18	24	30	3C	48	54	60	6C	78	84	90	9C	A8	B4
D	0	D	1A	27	34	41	4E	5B	68	75	82	8F	9C	A9	B6	C3
E	0	E	1C	2A	38	46	54	62	70	7E	8C	9A	A8	B6	C4	D2
F	0	F	1E	2D	3C	4B	5A	69	78	87	96	A5	B4	C3	D2	E1

Example (Hexadecimal Addition)

$$\begin{array}{r} B2 \\ +\ 3B \\ \hline =\ ? \end{array}$$

Starting with the right column and referring to the hexidecimal charts we obtain $2 + B = D$. In the left column $B + 3 = E$. Thus the result is ED_{16}.

Example (Hexadecimal Addition)

$$\begin{array}{r} \text{A4526} \\ +\ \text{EE01B} \\ \hline =\ ? \end{array}$$

Start with the right column, $6 + B = 11$. Here a 1 is "carried." In the next column the "carried" 1 plus the 2 and 1 in the column = 4. The middle column's addition yields 5 + 0 = 5. In the next column we obtain 4 + E = 12. A 1 is therefore "carried" into the last column. In it $A + E = 18$ and the "carried" 1, when added, produces 19.

$$\begin{array}{r} {\scriptstyle 1\ 1\ \ \ 1\ \ \ } \\ \text{A4526} \\ +\ \text{EE01B} \\ \hline =192541_{16}. \end{array}$$

Note the "carries."

Special Note

It has already been mentioned that programmers prefer to work in octal or hexadecimal arithmetic, as opposed to binary. To illustrate this point let us perform the same addition shown above, but this time in binary.

$$\begin{array}{r} {\scriptstyle 1\,1\,1\ \ \ \ \ 1\,1\ 1\,1\,1\,1\,1\ \ \ \ \ \ \ } \leftarrow\text{“carries”} \\ 1\ 0\ 1\ 0\ 0\ 1\ 0\ 0\ 0\ 1\ 0\ 1\ 0\ 0\ 1\ 0\ 0\ 1\ 1\ 0 \\ +\ 1\ 1\ 1\ 0\ 1\ 1\ 1\ 0\ 0\ 0\ 0\ 0\ 0\ 0\ 0\ 1\ 1\ 0\ 1\ 1 \\ \hline =1\ 1\ 0\ 0\ 1\ 0\ 0\ 1\ 0\ 0\ 1\ 0\ 1\ 0\ 1\ 0\ 0\ 0\ 0\ 0\ 1 \end{array}$$

Suppose, too, that we wanted to convert the hexadecimal and the binary results to "everyday" to decimal numbers. The binary to decimal conversion, via the polynomial expansion method is

$$\begin{aligned}110010010010101000001 &= 1 \times 2^0 + 0 \times 2^1 + 0 \times 2^2 + 0 \times 2^3 + \\ &\quad 0 \times 2^4 + 0 \times 2^5 + 1 \times 2^6 + 0 \times 2^7 + \\ &\quad 1 \times 2^8 + 0 \times 2^9 + 1 \times 2^{10} + 0 \times 2^{11} + \\ &\quad 0 \times 2^{12} + 1 \times 2^{13} + 0 \times 2^{14} + 0 \times 2^{15} + \\ &\quad 1 \times 2^{16} + 0 \times 2^{17} + 0 \times 2^{18} + 1 \times 2^{19} + \\ &\quad 1 \times 2^{20}. \\ &= 1 + 0 + 0 + 0 + 0 + 64 + 0 + 256 + \\ &\quad 0 + 1024 + 0 + 0 + 8192 + 0 + 0 + 65536 + \\ &\quad 0 + 0 + 524288 + 1048576. \\ &= 1{,}647{,}937_{10}.\end{aligned}$$

The hexadecimal to decimal conversion, via the polynomial expansion technique, is

$$\begin{aligned}192541 &= 1 \times 16^0 + 4 \times 16^1 + 5 \times 16^2 + 2 \times 16^3 + 9 \times 16^4 + 1 \times 16^5 \\ &= 1 + 64 + 1280 + 8192 + 589824 + 1048576 \\ &= 1{,}647{,}937.\end{aligned}$$

From the above it should be fairly obvious that binary arithmetic requires more steps than hexadecimal arithmetic. Also, a greater chance of error exists when performing binary calculations manually.

Hexadecimal Division and Subtraction

Base_{16} subtraction may be performed easily by using the addition table shown earlier. $E - 5$, for example, may be determined by looking across the 5 row and finding E. The column in which E appears is the required result; that is, 9_{16}.

Example (Hexadecimal Subtraction)

$$\begin{array}{r} \text{E7} \\ -\ 18 \\ \hline =\ ? \end{array}$$

In the right column we cannot subtract 8 from 7 until the left column is depleted by 1. Then $17 - 8 = F$. In the left column D remains; from it we subtract 1 to obtain C. The final result becomes CF_{16}.

Hexadecimal arithmetic may also be performed with the tables; 82 divided by *A*, for example, is *D*.

Example (Hexadecimal Division)

$$B \overline{)4A3} = ?$$

SOLUTION:

```
        6B
    B | 4A3
        42 . . . result of 6 times B.
        ---
        83 . . . the 8 comes from subtracting 42 from 4A.
        79 . . . or B times B.
        ---
         A
```

Result: $(6B + A/B)_{16}$.

COMPLEMENT ARITHMETIC

The finest digital computer cannot, as we do, deem a number negative when it sees a minus sign, nor can it deem a number positive when it sees a plus, but by the use of complement arithmetic the computer can, with just one addition or subtraction operation, add or subtract without the plus or minus sign. Two examples are given to demonstrate how, by the use of complements and the "end-around carry," we may subtract or add in the digital computer.

1. Add zeros to the subtrahend, if necessary, so that the number of characters in it equals the number of characters in the minuend.
2. Create the *ones complement* of the subtrahend. (In binary the ones complement is simply an interchange of 1's and 0's; for example, the ones complement of 1000110 is 0111001.)
3. Add the complement to the minuend. The result should consist of n characters, where n = the number of characters in the minuend. When an "extra" character exists, remove it and add it to the 2^0 column of the result.

Example. Subtract 1011 from 101101.

Step 1. Change 1011 to 001011.
Step 2. The complement = 110100.

Step 3. Adding the minuend and the complemented subtrahend yields

```
                      101101
                  +   110100
                  ----------
                  =  1100001
end-around carry    └──────→ 1
                  ----------
                      100010 = result.
```

Example. Subtract 1011 from 100111.

Step 1. Change 1011 to 001011.
Step 2. The complement = 110100.
Step 3. Adding the minuend and complemented subtrahend yields

```
                      100111
                    + 110100
                  ----------
                     1011011
end-around carry    └──────→ 1
                  ----------
                  =   011100 = result.
```

11

COMPUTER OUTPUT MICROFILM, MICROFILM CAMERAS AND STANDARD MICROFORMS

A highly cost-effective means of recording and storing large quantities of analog or digitally processed transducer or sensor data is by microfilm.

Measurement listings, graphs, maps, and images or photos may be reproduced on film. When computer listings are so recorded, a 50-to-1 volume/weight savings can be realized. Computers that output microfilm (as opposed to line-print pages) frequently provide cost savings in terms of computer time of approximately six-to-one. Computer output microfilm (COM) systems are not hampered by slow line printers or computer output plotters. Instead, final listing or annotated graphics data are routed to a cathode ray tube (CRT) and then photographed. Thirty or more pages or graphs can be placed on film per second. In no way can a computer line-printer typewriter type that fast or can a computer output plotter generate individual plots. A COM system is especially useful when large quantities of data must be printed, stored, and distributed. In addition to COM, rotary and planetary cameras can photograph charts, maps, photos, imagery, reports, and other documents of any type and size. Hundreds of original data products, when stored on film, eliminate the problems of high volume, high shipping or mailing costs, and the high cost of document duplication when many copies are required.

Final microfilm, whether computer-generated or photographed by the planetary or rotary camera, may be viewed on film readers. Many are equipped with print options so that film images can be converted to hard copy prints. Conversion of rolls of film to high-quality bond-paper copy is also possible.

Discussed in the paragraphs that follow are the fundamentals of microfilm technology; that is, the basic computer output microfilm system, COM advantages and disadvantages, microfilm cameras, and four common microforms. The basic computer output microfilm system is illustrated in Figure 1. The logic/conversion section performs four primary functions:

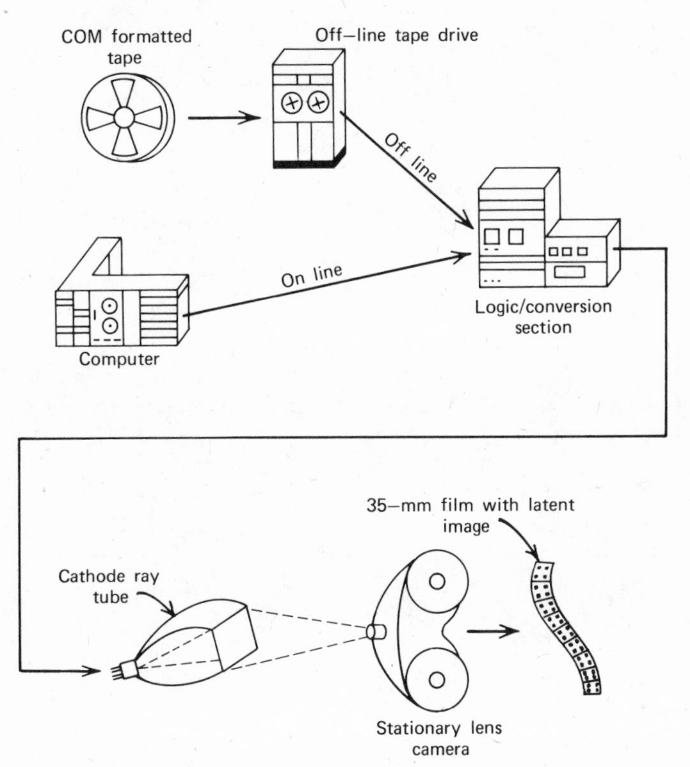

Figure 1

1. It determines what data are to be routed to film and in what size and position data are to be recorded onto each film frame.
2. It controls coding and error detection.
3. It translates incoming digital data (from tape or computer) to a

predetermined analog signal pattern by computer software or character/vector generating circuits.

4. It deflects analog beams (which represent numbers, vectors, or letters) to desired CRT screen locations.

The CRT serves to project (or stroke) electrons onto the tube face to produce visible images. Images may also be photographed by the stationary lens camera that advances film by tape or computer command. Film is usually conventional silver. 35-mm film is processed, copied, and read as shown in Figure 2.

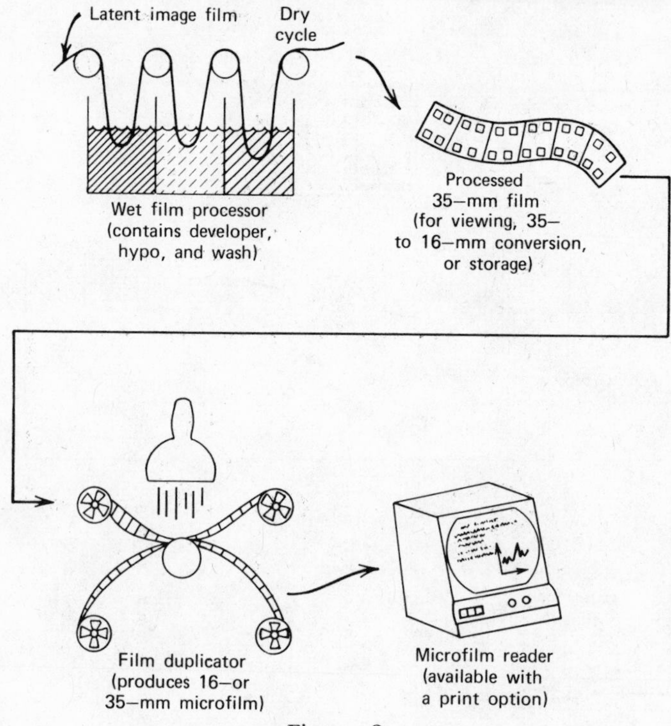

Figure 2

The alphanumerics or alphanumerics/graphics computer output microfilm system offers many advantages and some disadvantages.

COM Advantages (partial list; typical industry numbers shown)

1. Print time ratio (vs computer line printer) is 10:1.

2. Computer time savings ratio is 8:1.
3. Physical volume/weight is 1/50th that of line-printer paper.
4. Ease of storage.
5. Film-packing density for certain applications is 1000 times that of magnetic data tape.
6. Materials cost one-sixth of line printer expendables.
7. Low film-duplication cost.
8. COM effectively records data at computer speeds. Line printers do not.
9. Graphics displays including computer forms creation, are possible.

COM Disadvantages

1. Stored data is unavailable in real or near real time.
2. Requires programmer support.
3. Large capital investment is required:
    - readers, reader/printers.
    - processors.
    - duplicators.
    - microforms generation hardware.
4. Standard COM systems produce outputs that cannot be annotated.
5. Per page (frame) costs are acceptable only under high volume conditions.
6. Users must be trained.
7. Hardware failures cause backlogs of a large number of pages.

Generally speaking, COM is recommended when high volumes of data are to be generated, when a 1- to 2-day turnaround time is acceptable, and when users prefer working with film readers, as opposed to handling large quantities of printer paper.

MICROFILM CAMERAS

In addition to the microfilm camera installed in COM systems are two others—the planetary and the rotary (Figure 3). The planetary

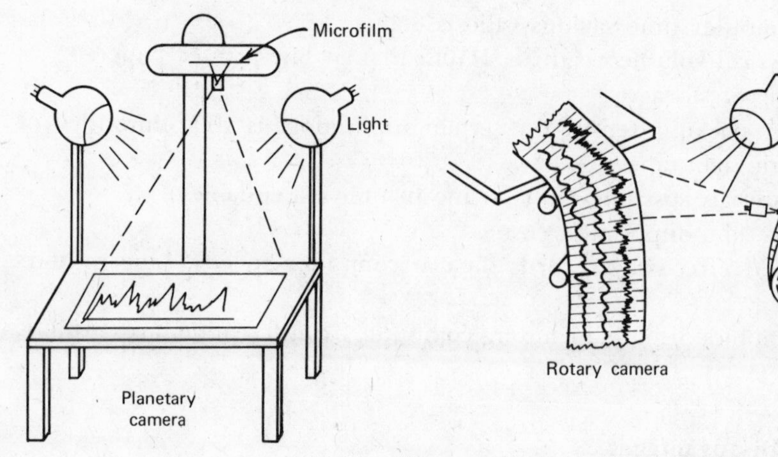

Figure 3

cameras are designed to photograph large-scale documents, drawings, and blueprints. The rotary camera films large quantities of computer line-print paper, reports, small charts, and roll data like that generated by the analog pen recorder or oscillograph.

COMMON MICROFORMS

Among the common types of microfilm are 35-mm and 16-mm roll film, aperture cards, microfiche, and film jackets. Each microform type contains reduced film images. Reduction ratios, paper-to-film, or CRT-to-film generally fall within the 2 to 300X range.

Roll film, wound on a spool, comes in 16 or 35mm and may be obtained in 100-, 400-, or 1000-ft lengths. Image orientation on film has the *cine* or *comic* format, and photographic reduction is usually 10 or 40X (Figure 4).

Microfiche, usually 4 × 6 in., contains cathode ray tube images recorded in rows or columns. Fiche, generated by a Universal camera, is generally equivalent to 224 pages of data (Figure 5). Each fiche may be read on a microfiche reader, many of which have print options. Easy to store and requiring only 2% of the storage space for an equivalent quantity of line-printer data, 1 lb of fiche equals more than 250 lb of paper.

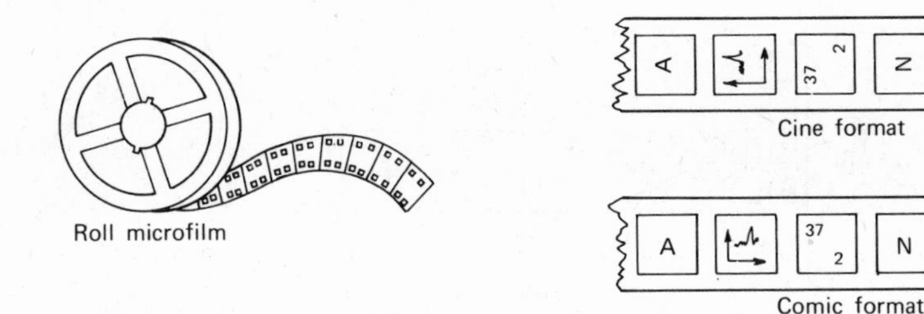

Figure 4

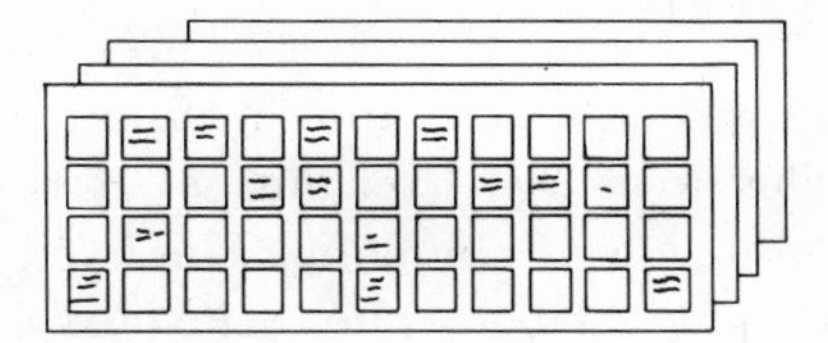

Figure 5

Many magazines, books, and scientific reports are available on microfiche. NASA earth-resources experiment results are an example. Although more expensive to generate than roll film, microfiche is suitable for extensive distribution and is often preferred by librarians and users of scientific reports.

Roll-film data, when placed in cassettes, is generally five times heavier than an equivalent quantity of data on fiche.

Aperture cards, each about the size of a keypunch card, have one or more rectangular holes for housing film frames. A completely filled blueprint or photo cabinet can be reduced to a small quantity of cards. Aperture cards, like microfiche, are ideal when many document copies must be distributed (Figure 6).

Film jackets contain inserted frames of microfilm. One application is patient medical records. Progress reports and laboratory tests can be inserted, as frames of film, as the case history expands. (Figure 7).

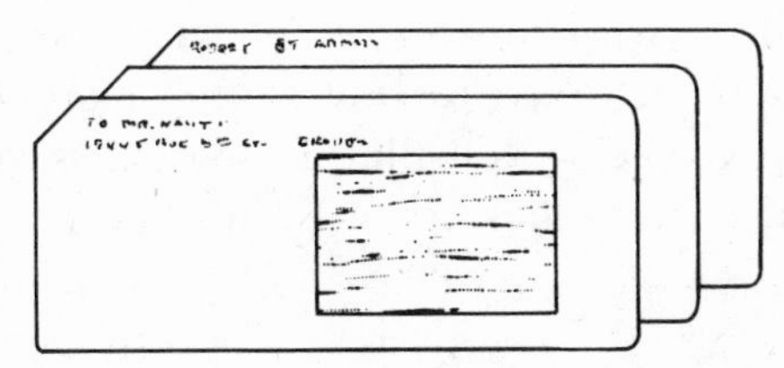

Figure 6

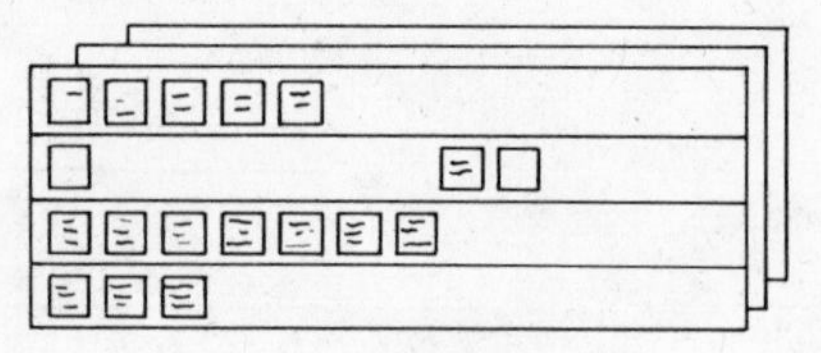

Figure 7

FORM PRODUCTION

An advantage of the computer output microfilm system worthy of discussion is its ability to merge fixed images (forms) with computer data during the production of microfilm. The fixed image, on a form slide, may be a letterhead, logo, photo, invoice grid, map, or any other item suitable for combination with computer-generated graphs or listings.

The form slide is an interchangeable glass, acetate, or polyester slide which contains a fixed image that may be projected for superimposition with computer-generated data. Form projection can be such that under program control any one of a number of forms can be flashed or a specific number of forms can be flashed in sequence with COM film frames. The fiirst mode is referred to as random; the second is sequential. Also possible is single-form projection in which only one form is used for all output data.

An obvious advantage of the form-flashing process is its capability of producing a variety of business or technical documents without the expense of preprinted computer paper.

COM ROLL FILM-TO-PAPER PRODUCTION

Following the review of microfilm by use of a roll film reader, it is often necessary to convert film to 8½ × 11 in. sheets of paper or vellum. This may be achieved by one of the many available electrostatic film-to-paper copiers. Roll film, as a rule, is converted to roll paper which in turn is cut into 8½ × 11 in. sheets at the end of each frame.

12

BASICS OF DYNAMIC ANALYSIS

Differences in the optical properties of earth-resource features may be detected and measured by *X-Y* plots that exhibit sensor spectral response versus frequency. Analysis of acoustic data, whether from land, sea, or air, usually requires the production of acoustic energy versus frequency displays. This is also the case when the effects of vibration on a component, subassembly, or structure are to be studied.

Useful dynamic analysis techniques for the study of physical and optical data are the power spectral density function, autocorrelation, cross correlation, and the probability density function. All are discussed in the pages that follow. Performance of dynamic analysis may be achieved by analog, digital, or hybrid (i.e., analog/digital) techniques. Choice of the proper analysis system is governed by several factors:

1. *Quantity of data*. Computer cost effectiveness is often related to the number of analyses to be performed.
2. *Filtering requirements.* Power spectral density analysis, for example, requires bandpass filtering. Analog filters are frequently difficult and costly to duplicate with computer software.
3. *Output plot format*. Analog *X-Y* plotters and computer output plotters produce high-quality plots but require substantial plotting time. The computer graphics cathode ray tube produces plots rapidly

but requires computer output microfilm or a CRT-to-photo product system to produce hardcopy.

Analog versus digital arguments are usually settled by specifying the quantity and format of input data and desired output products. These factors, when weighed against available hardware and software, enable the digital versus analog or hybrid decision to be made.

Although the sections that follow are specifically addressed to analog and hybrid data processing, note that the principles and concepts apply also to computer programs that perform dynamic analysis operations.

PHYSICAL DATA

For the most part, data that are recorded by the transducer and remote sensor fall into two categories, deterministic and random.

Deterministic data are those that repeat themselves identically each time a test or experiment is repeated. Random data do not. Examples of deterministic data include the following:

Sinusoidal periodic data. Monitoring the waveform of the electrical signal present in home wall outlets reveals an identical 60-cycle curve each time monitoring with a scope is performed.

Transient nonperiodic data. The temperature response by a light bulb turned off (Figure 1). The sound made by a door slammed with

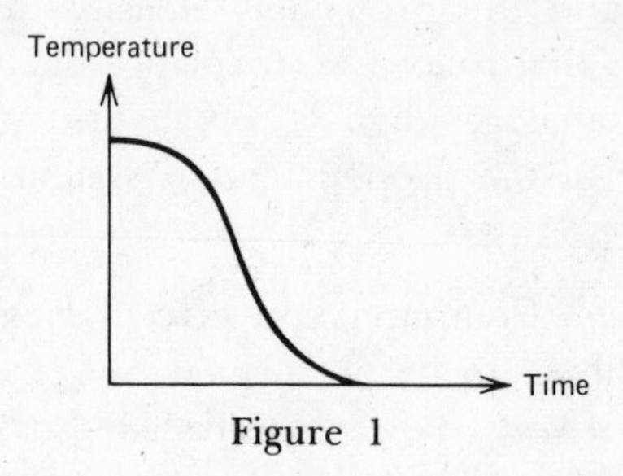

Figure 1

a fixed force (Figure 2). The vibratory action of a spring expanded (a fixed distance) and let loose (Figure 3).

Complex-periodic data. A data signal containing one basic frequency component plus that frequency's harmonics. (A harmonic is an inte-

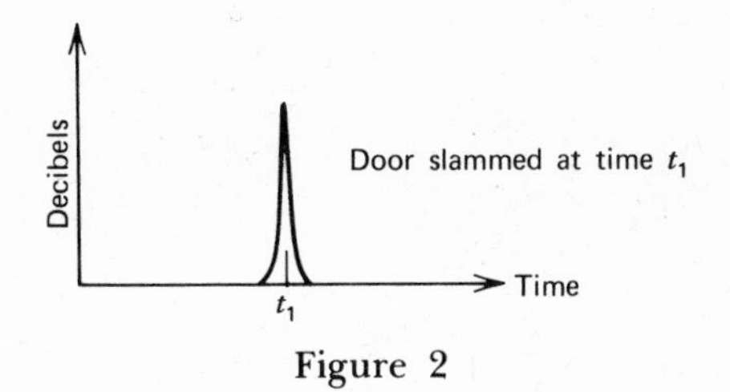

Figure 2

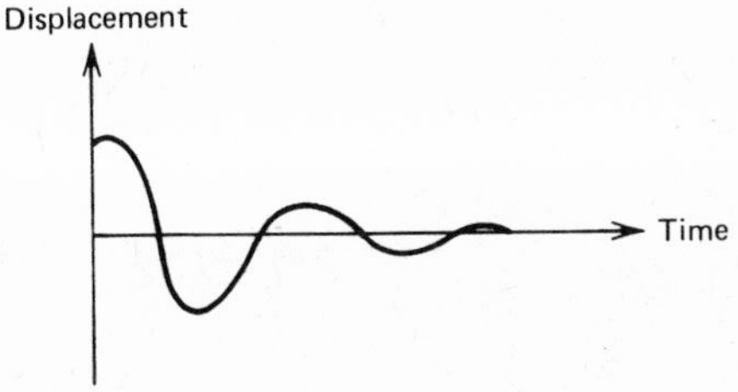

Displacement

Time

Figure 3

gral multiple of the basic frequency). The vibrations associated with an automobile engine usually contain complex periodic components.

Examples of random data would occur if at the same time each day we released like amounts of smoke into the wind and photographed the results (Figure 4); if each day we released a mouse in a New York City subway car and recorded voice reactions (Figure 5); and if we recorded on an oscillograph recorder the voltage waveforms pro-

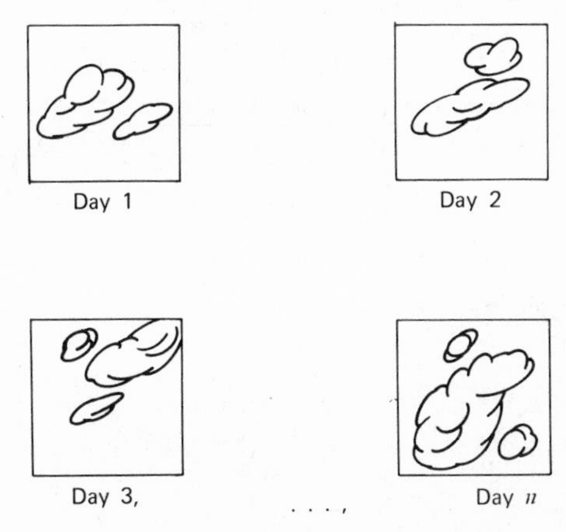

Figure 4

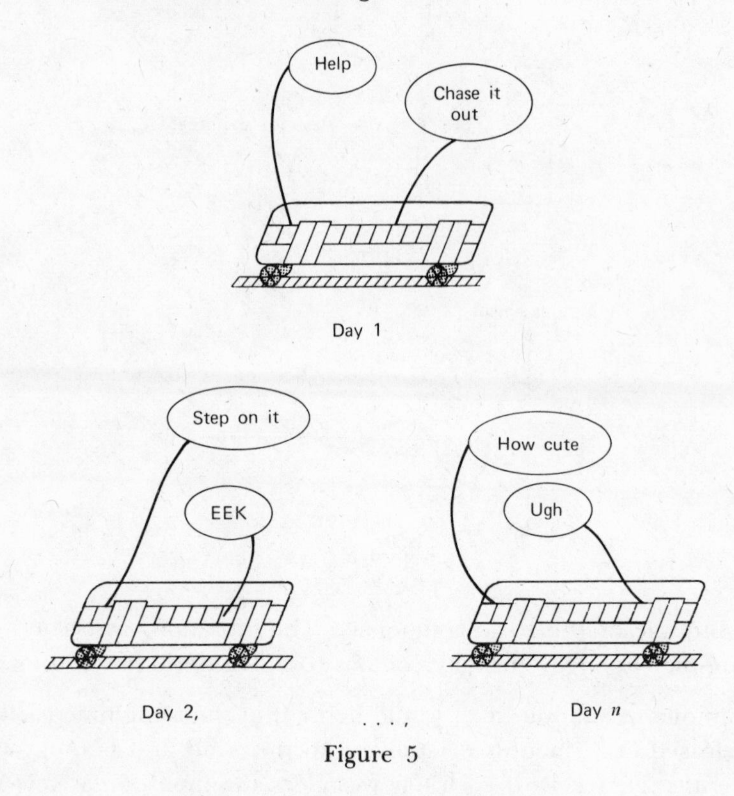

Figure 5

duced by several identically constructed electrical noise generators* (Figure 6).

Random data occur often. Unfortunately, they cannot be described by an explicit mathematical equation. Analysis of random data, however, is not difficult to perform. When random data are *ergodic,* as great amounts of data are, analysis may be performed with standard dynamic analysis hardware or software, but before the term ergodic is defined, a definition of stationary random data must be stated.

Stationary Random Data

Consider a flag experiencing the effects of wind (a random process; Figure 7). For each day let us draw a graph of $h(t)$ versus time (Figure

*A noise generator is a piece of laboratory equipment that generates a voltage signal which, from moment to moment, is never identical in form.

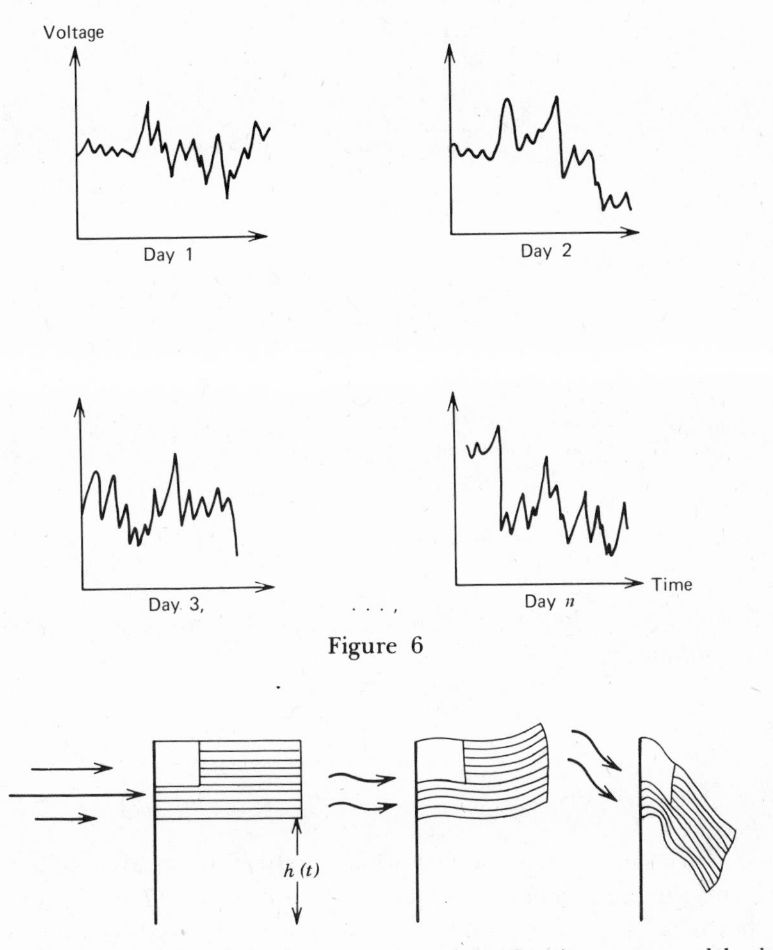

Figure 6

Figure 7. The height of the lower corner of the flag is represented by $h(t)$ with respect to time.

8). Graphs like this are commonly known as *time histories*. For the n-day period let us calculate the *mean value* at time T_1, where T_1 represents 9:30 a.m.:

$$\text{mean value} = \frac{\underset{\text{at } T_1}{h(t)_1} + \underset{\text{at } T_1}{h(t)_2} + \ldots + \underset{\text{at } T_1}{h(t)_n}}{n}$$

Now consider another calculation, sometimes referred to as the joint moment between values at two different times: T_1 and $(T_1 + X)$, where

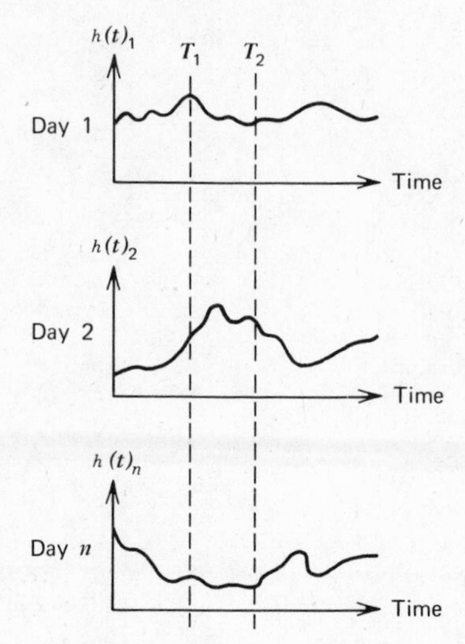

Figure 8. The height of the corner of the flag during day 1 is $h(t)_1$.

X is some fixed time later. This calculation is known as the *autocorrelation function A*:

$$A = \frac{\underset{\text{at } T_1}{h(t)_1} \times \underset{\text{at } T_2}{h(t)_1} + \underset{\text{at } T_1}{h(t)_2} \times \underset{\text{at } T_2}{h(t)_2} + \ldots + \underset{\text{at } T_1}{h(t)_n} \times \underset{\text{at } T_2}{h(t)_n}}{n}$$

where $T_2 = T_1 + X$.

Data are said to be stationary when neither the mean value nor the autocorrelation functions vary as T_1 changes (i.e., if T_1 changes from 9:30 a.m. to 9:31 a.m., 9:32 a.m., and so on). When the mean value remains constant as T_1 varies but the autocorrelation function varies as X varies, data under observation are said to be weakly stationary.

Vibration data, undersea acoustic data, satellite remote sensor data, or any other form of "dynamic" data that are either stationary or weakly stationary may be processed by standard dynamic analysis techniques with no highly significant error. Nonstationary data, on the other hand, require special analog or digital data processing techniques and the validity of final plots and listings are not always trustworthy. When data are stationary, data handling is not too complex. When data are ergodic, the data handling problem is even further simplified.

Ergodic Random Data

Simply stated, if a random process (group of data) is stationary and the mean value and autocorrelation functions remain constant for each sample in the data group, the process is deemed ergodic:

$$\text{mean value for the } n\text{th sample} = \lim_{T\to\infty} \frac{1}{T} \int_0^T h_n(t)\, dt$$

$$\text{autocorrelation} = \lim_{T\to\infty} \frac{1}{T} \int_0^T h_n(t) \times h_n\,(t + X)\, dt$$

Once data from the transducer or sensor are proved ergodic, final test data processing can be produced from a single sample. In some cases the analysis of many seconds or minutes of data can be reduced to processing only one sample a few seconds long. This factor is especially important when the final data user must pay for analog or digital machine time. Almost always the smaller the quantity of data, the lower the data processing cost.

In actual practice a majority of all stationary data are ergodic. This is especially true when data acquired during vibration testing or in the recording of acoustic energy are studied.

Dynamic analysis techniques, as we shall see in the forthcoming sections, offer means of describing data physically. Although the study of dynamic analysis methods is a complex one from the standpoint of mathematics, major concepts may be easily understood by using of relatively simple physical examples.

The sections that follow list mathematical equations but also describe their physical meaning. Concepts are important; the equations are included chiefly as reference material for those readers who require more than a basic knowledge of dynamic analysis techniques.

DYNAMIC ANALYSIS—A VALUABLE MEANS OF DESCRIBING PHYSICAL PHENOMENA

When an object generates sound, that sound contains frequency components, some more dominant than others. Differences in sound between two sound-generating features are due to the strength differences in detected frequency components. Plotting detected acoustic response versus frequency yields different plots for different objects (Figure 9). When one object's response curve (signature) is unique

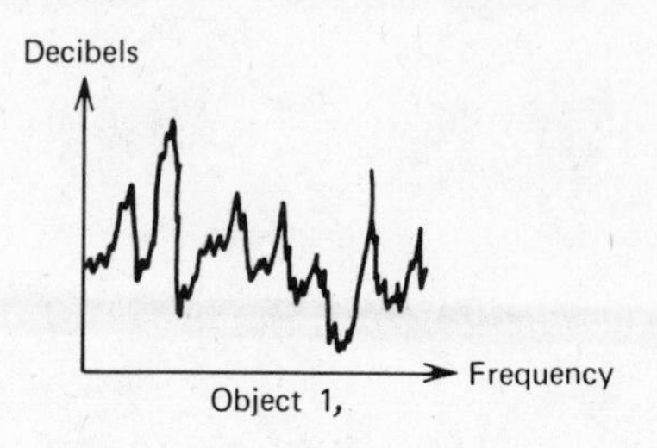

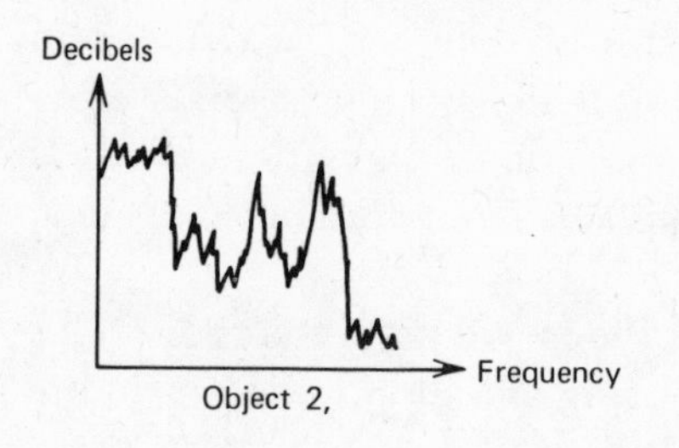

. . . ,

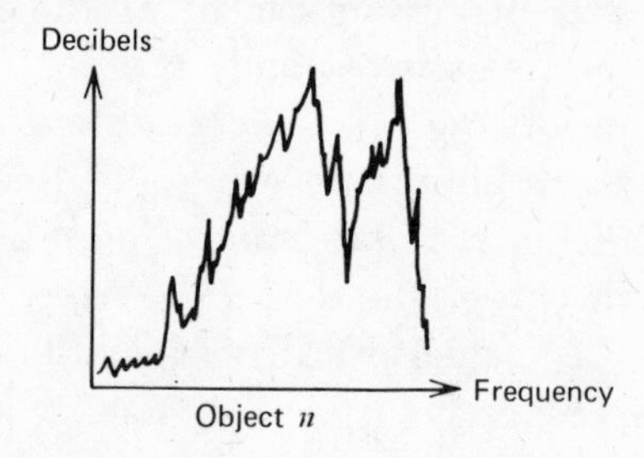

Figure 9

with respect to other known radiating objects in a given geographical region, the presence of that feature may be inferred each time a spectrum analyzer displays that signature. (This concept is important in antisubmarine warfare.)

Reflected and emitted energy patterns for different farm crops, detected by the multispectral scanner, differ in frequency (or wavelength) composition. When different ground vegetation, trees, and crops yield unique spectral signatures, these signatures serve as the basis for vegetation-feature monitoring by the remote sensor. Changes in a feature's response curve often relate to drought, insect infestation, or lack of fertilizer.

Both analog dynamic analysis hardware and the properly programmed digital computer play major roles in the study of acoustic data (detected by a microphone or underwater hydrophone), in evaluating optical data (obtained by the multispectral scanner), and in the analysis of vibration data recorded by *in situ* transducers.

The need for extensive vibration studies is especially important, for example, during the manufacture of ground, air, and space vehicles. Let us examine first the factors that justify this statement and start by trying a simple experiment.

Hold the edge of a ruler, stick, or metal rod between your thumb and index finger and shake the "structure" slowly. What happens to the top end? Now shake rapidly; observe the effect (Figure 10). Large

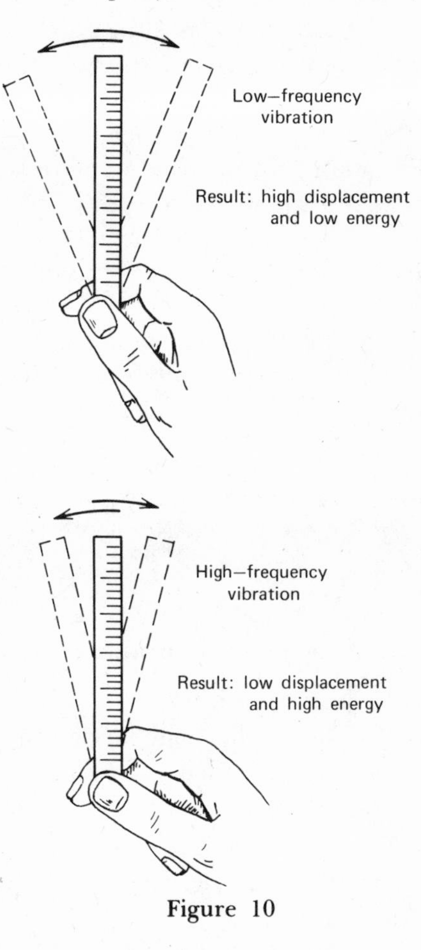

Figure 10

displacements and low energy are usually associated with low-frequency vibration, whereas the reverse is true when high-frequency vibration occurs.

It can also be demonstrated that small items of solid construction (nuts, bolts, resistors, and switches) tend to be unresponsive to rapid vibration. Large items, on the other hand, tend to respond to both classifications of vibration.

Vibrations Can Destroy

When subjected to a vibration of a harmful magnitude and frequency, a specimen will *literally tear itself to pieces*. The purpose of a vibration test program is to survey structures to determine by controlled methods whether they will function properly and without failure during their intended use time. When one or more parts of a structure (e.g., plane, car, or train) experience severe vibration affects, they can be damaged or break loose. It is also possible that oscillations in one part of a structure can by energy transmission cause problems in another.

Why and how do structures and components fail when subjected to certain kinds of physical or acoustic vibration? The existence of resonance is a key factor.

RESONANCE. Resonance exists in a system when, with constant excitation (e.g., vibration or acoustic energy), any change in the excitation frequency produces a decrease in the response experienced by that system.

Often excitation at a resonant frequency for an extended time can cause system response to increase. Two additional factors must also be considered. (Figure 11)

LIMIT OF ELASTICITY. Rigid bodies may be deformed up to this limit and will return to their original condition when the deforming force is removed.

FATIGUE LIMIT. A measure of the number of times a rigid body may be deformed elastically up to the limit at which the body structurally fails.

These definitions, when combined, tell us that if a vehicle, assembly, or component experiences excitation at its resonant frequency for an extended time it could be deformed or fatigued to failure (Figure 12).

Dynamic analysis plays a major role in the area of underwater re-

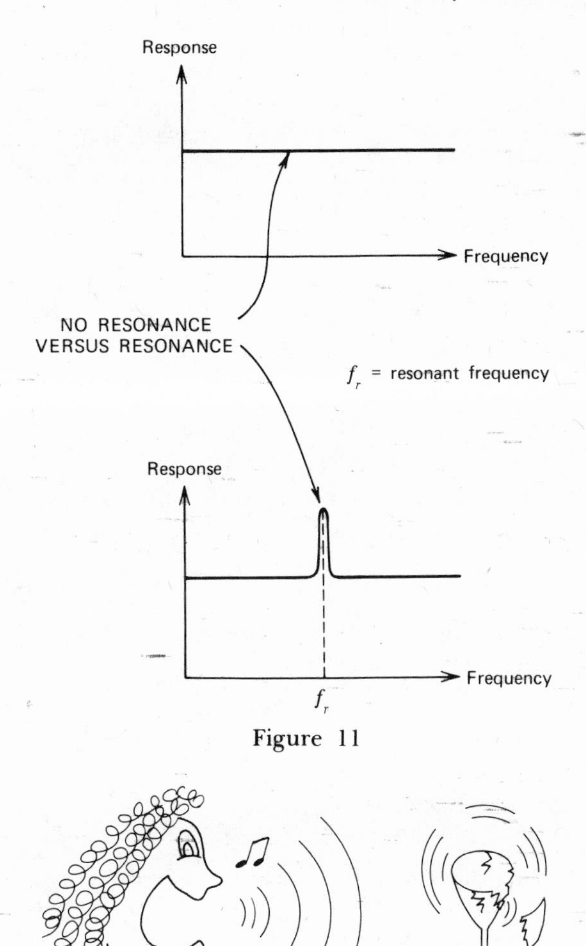

Figure 11

Figure 12

search. Analysis of sonic and acoustic signals can reveal factors related to, for example, submarine type and position, sea-floor surface details, types of marine life, and water surface traffic (Figure 13). Remote sensor data from the multispectral scanner, when analyzed to produce response versus wavelength plots, yields data valuable in studies that are related to agriculture, forestry, range resources, landform surveys, water resources, and land-use mapping.

Clearly, the description of data in terms of its frequency composi-

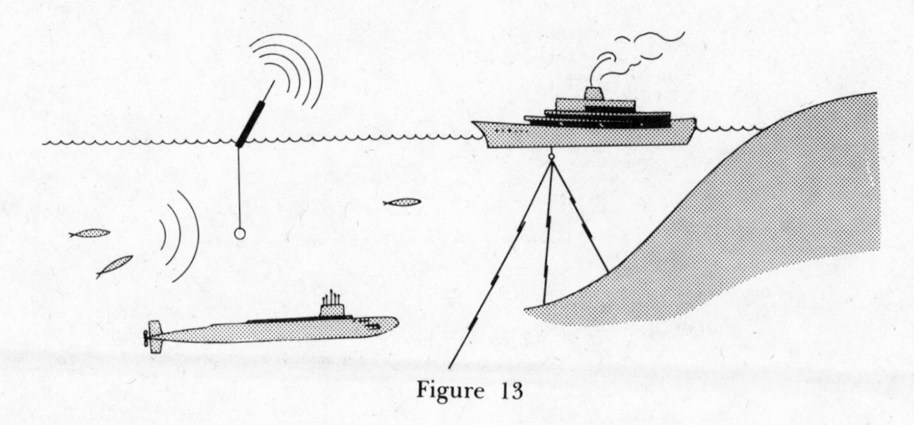

Figure 13

tion is an integral part of environmental data handling. One of the most important and revealing types of analysis that the test data processing facility can perform is spectral density analysis. Response plots generally have one of these formats: amplitude versus frequency, power versus frequency, or amplitude versus wavelength. Each relates to the other. The section that follows discusses the power spectral density function and typical analog analysis techniques.

POWER SPECTRAL DENSITY ANALYSIS

The power spectral density (PSD) function, like the amplitude spectral density (ASD) function, reveals the frequency composition of data, which in turn, yields important relations in regard to the characteristics of the physical system or earth feature being monitored.

NOTE. The PSD plot is generally used in acoustic and vibration analysis surveys. The ASD plot is similar except that no squaring of sensor or transducer output voltage occurs in the ASD plotting system.

Both PSD and ASD plotting systems present the physical frequency characteristics of a system or earth feature. Let us, for example, illustrate a spectral density plot generated from sounds recorded in a kitchen (Figure 14). A PSD plot of the electrical data signal in Figure 14 is shown in Figure 15.

PSD and ASD plots are valuable tools in several technical and scien-

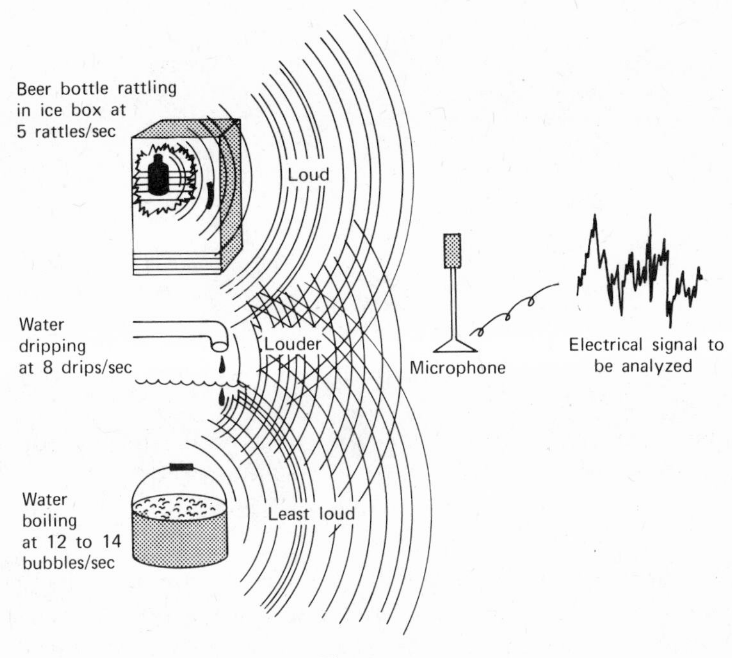

Figure 14

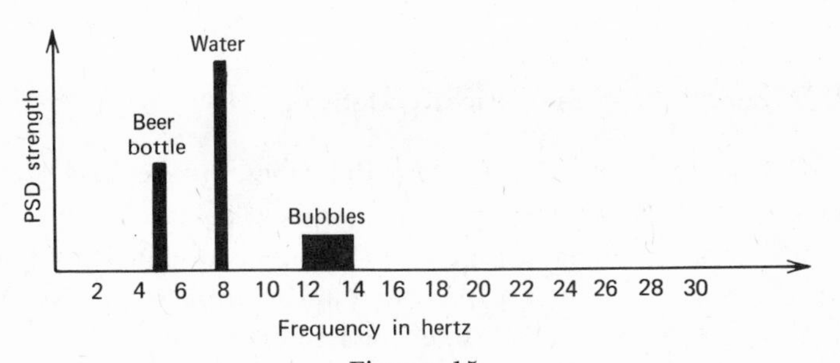

Figure 15

tific fields. During vibration testing of structures, for example, spectral density plots give engineers a detailed look at the response characteristics of the body under test. When final plots show resonance characteristics, the possibility exists that problems will occur during the structure's intended life time (Figure 16). In undersea studies the marine biologist and the oceanographer find spectral density plots useful in detecting and classifying certain forms of marine life or man-made structures. Because each dynamic (active) physical system

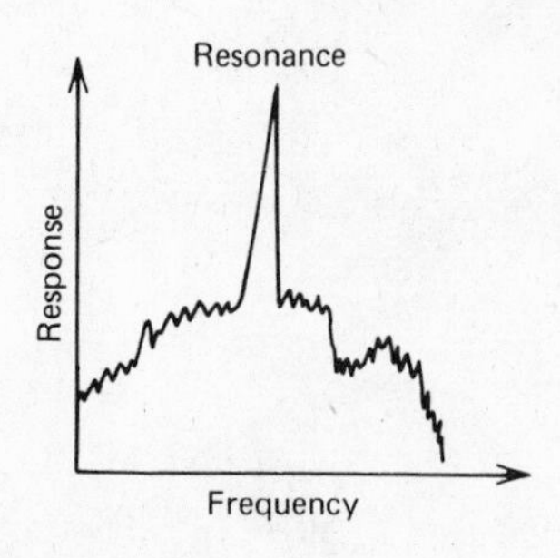

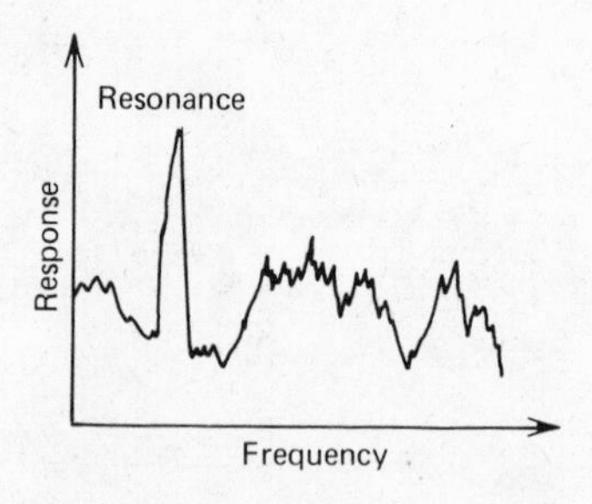

Figure 16

(living or man-made) is unique, spectral density plots differ from one feature to another.

Similarly, in studying earth-feature data from the multispectral scanner (discussed in Chapter 2), plots of optical response versus frequency are used to detect and monitor the status of crops, water, and forests.

Unique active physical systems have different response characteristics. The spectral density plot, generated from sensor or transducer data, is a widely used tool in evaluating the frequency content of each pertinent data signal (Figure 17).

The PSD Function Mathematically Defined

The PSD function for random data describes the general frequency composition of data in terms of mean-square spectral densities. The mean value in a frequency range f to $(f + \Delta f)$ is obtained by bandpass filtering the transducer or sensor signal and computing the average of the squared output from that bandpass filter.

$$\mathrm{PSD} = \lim_{T\to\infty} \frac{1}{T} \int_0^T v^2(t, f \to (f + \Delta f))\, dt$$

where $v(t, f \to (f + \Delta f))$ is the portion of the signal $v(t)$ in the frequency range from f to $(f + \Delta f)$.

When data are random and stationary, the PSD function may be estimated by

$$\mathrm{PSD} = \frac{1}{BT} \int_0^T v^2(t)\, dt$$

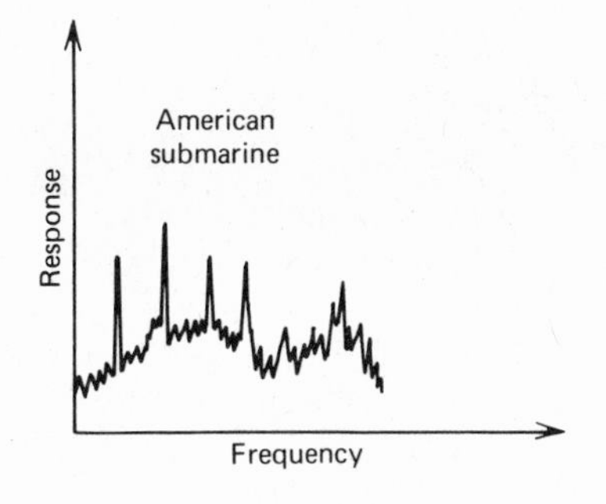

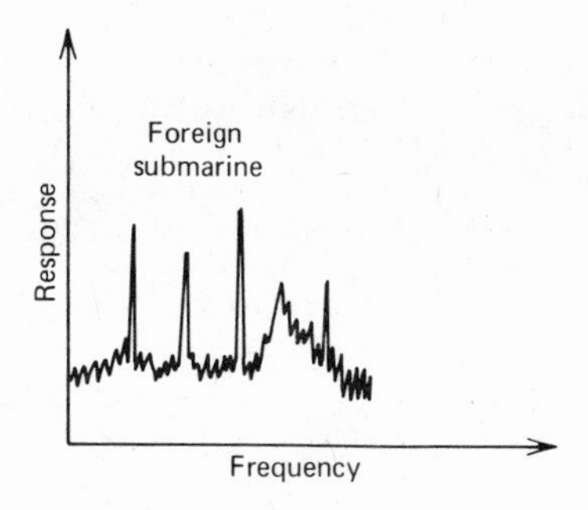

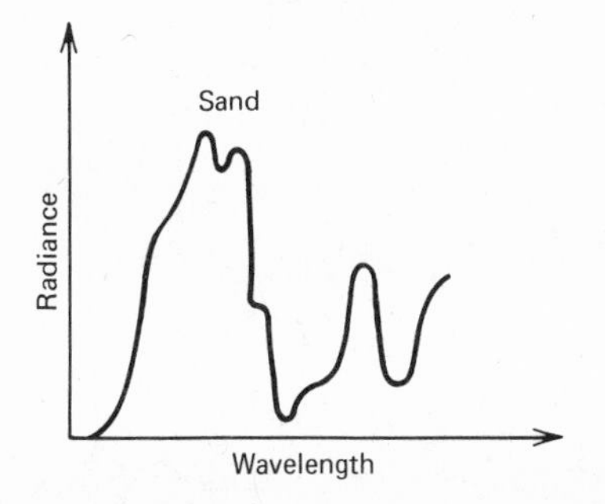

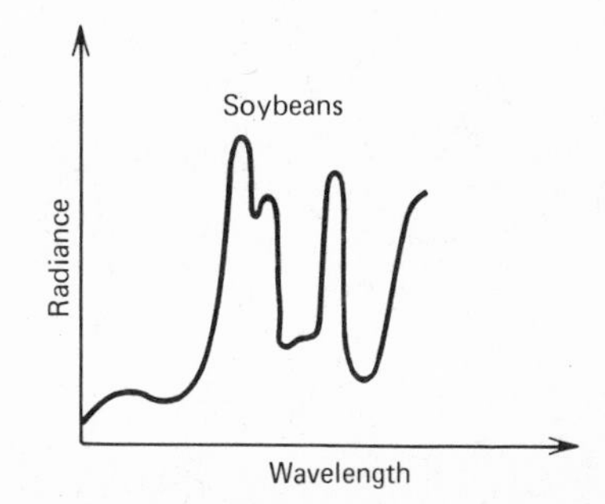

Figure 17

where $v(t)$ is the portion of the data sample passed by a narrow bandpass filter with a bandwidth of B Hz.

In simpler terms, this equation calls for frequency-filtering the data signal $v(t)$ with a bandpass filter that is B Hz wide, squaring the filter output to create $v^2(t)$, then averaging the squared value over the sampling time (0 to T), and dividing the mean square output by the value B.

The PSD function is frequently difficult to understand physically by looking only at equations. Because of this factor, the sections that follow immediately include two simplified physical examples that describe the manner in which a PSD plot is generated. In each the PSD function is defined as highly analogous to the relationship v^2/B.

Example 1. Backyard Noises. Assume that a microphone records backyard sounds with the intensity (i.e., loudness) of individual sounds analogous to the number of volts each causes the microphone to generate. Assume that three sound generators cause the following:

1. Water dripping from a hose:	1 drip/sec—microphone output = 1 V
2. Dogs barking:	3 barks/sec—microphone output = 4 V
3. Birds chirping:	6 chirps/sec—microphone output = 2 V

This example is illustrated in Figure 18. Now, if the selected bandpass

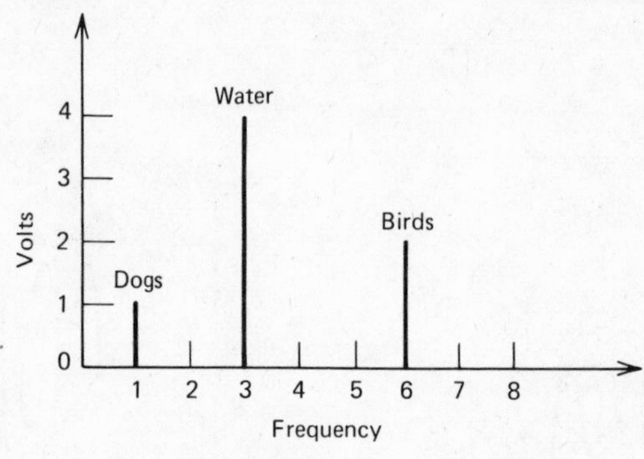

Figure 18

filter is 3 Hz wide, PSD analysis proceeds by first placing it (shown shaded) on frequency = 1 Hz, then on 2 Hz, 3 Hz, and so on (Figure 19). As the bandpass filter (analog or digital) is sequentially stepped down the x axis, the analysis hardware (or computer software) looks to see if any signal is observed in the bandpassed (shaded) region. When sensor or transducer response energy is detected, the calculation v^2/B is performed with the result plotted on a PSD plot.

More specifically, the PSD system plots calculated v^2/B values at each center frequency point of the bandpass filter. Note, for example, that when B is placed (or centered) on 3 Hz, the system becomes capable of detecting energy in the 1.5 to 4.5 Hz band; the band is B, or 3 Hz wide.

No matter what the center value, B is always 3. If an x axis = 7, energy in the 5.5 to 8.5 Hz band is observed. (Figure 20).

Example 2. Kitchen Sounds. Assume that the backyard microphone has been moved into a kitchen with four physical/microphone output characteristics:

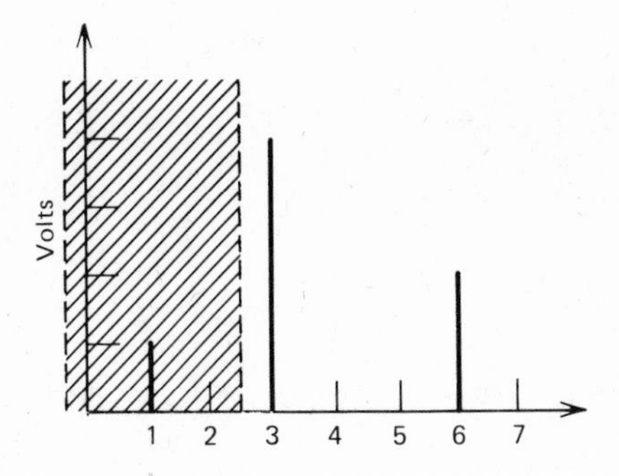

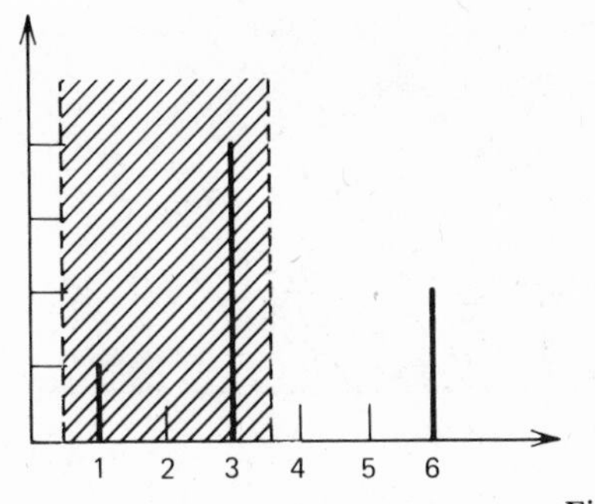

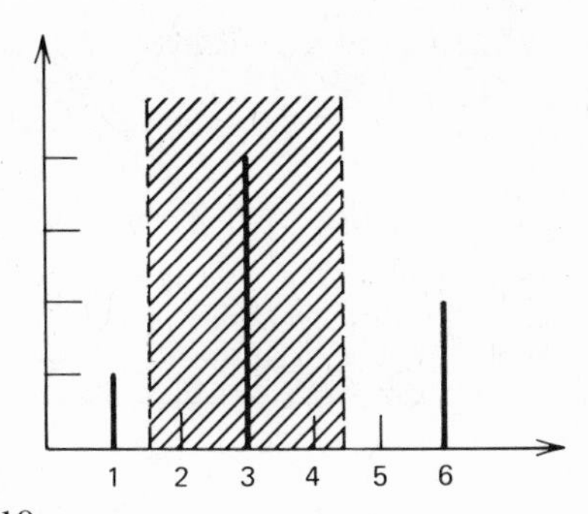

Figure 19

1. Window rattling — 1 rattle/sec at 4 V
2. Popcorn popping — 3 pops/sec at 3 V
3. Water boiling — 5 bubbles/sec at 1 V
4. Blender chopping nuts — 8 chops/sec at 2 V

This system is illustrated in Figure 21. Performing the v^2/B calculation for $B = 5$ Hz and $B = 3$ Hz, at x axis = 1, 2, . . . , 8 yields the following:

x axis	Bandpass Range (in Hz)	Volts Observed and v^2/B for $B = 5$		Volts Observed and v^2/B for $B = 3$	
1	0 to 3.5	7	9.8	4	5.3
2	0 to 4.5	7	9.8	7	16.3
3	0.5 to 5.5	8	12.8	3	3.0
4	1.5 to 6.5	4	3.2	4	5.3
5	2.5 to 7.5	4	3.2	1	0.3
6	3.5 to 8.5	3	1.8	1	0.3
7	4.5 to 9.5	3	1.8	2	1.3
8	5.5 to 10.5	2	0.8	2	1.3

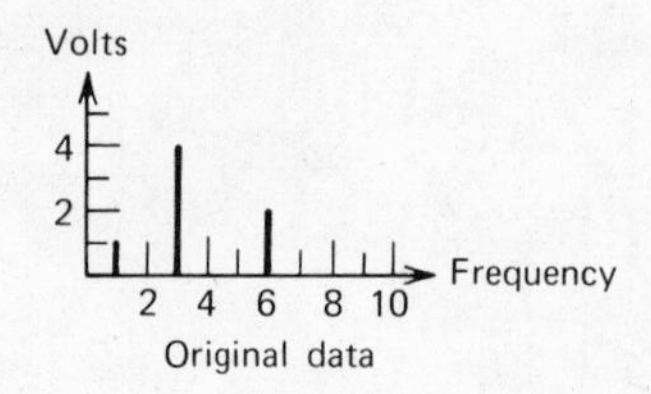

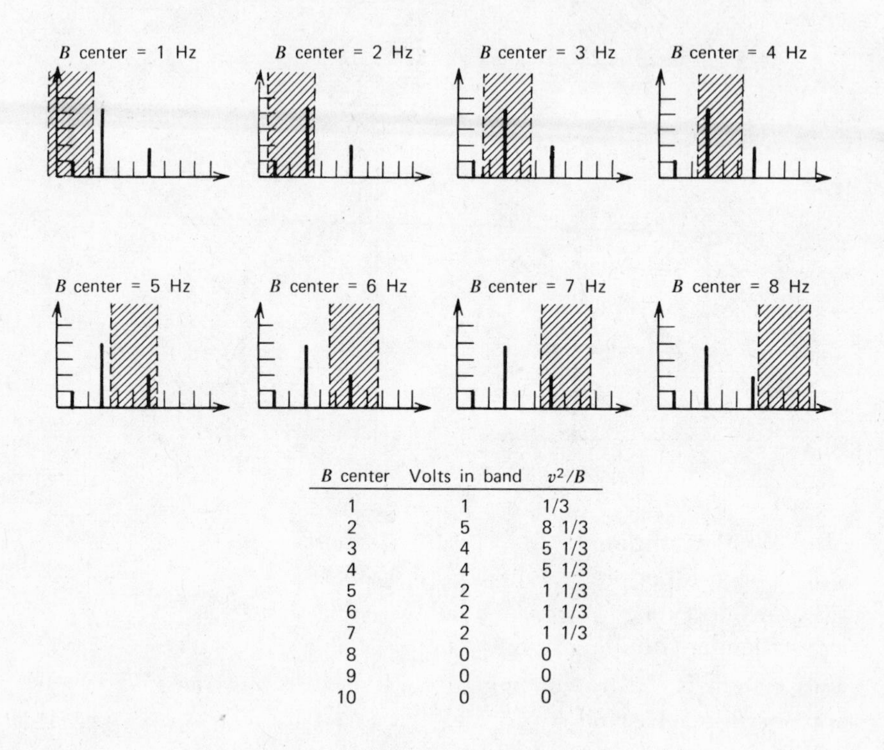

B center	Volts in band	v^2/B
1	1	1/3
2	5	8 1/3
3	4	5 1/3
4	4	5 1/3
5	2	1 1/3
6	2	1 1/3
7	2	1 1/3
8	0	0
9	0	0
10	0	0

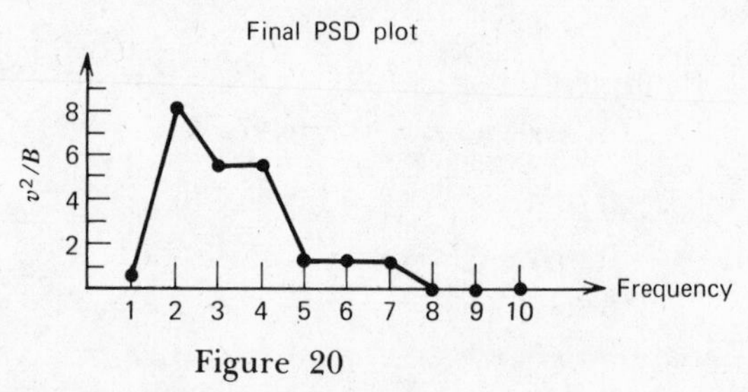

Figure 20

Final PSD plots for $B = 5$, $B = 3$, and $B = 1$ are shown in Figure 22. Note the differences as B decreases.

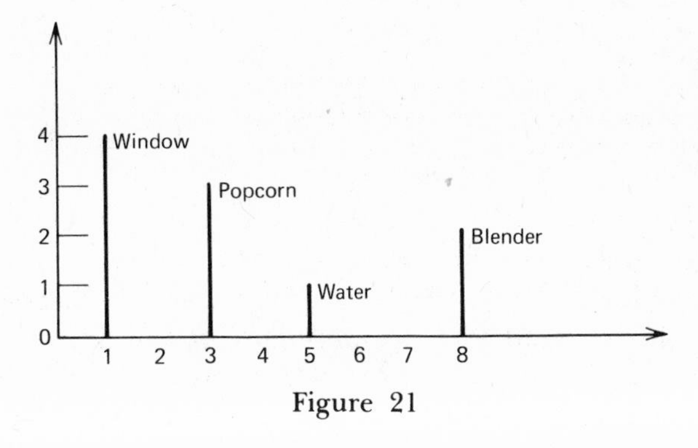

Figure 21

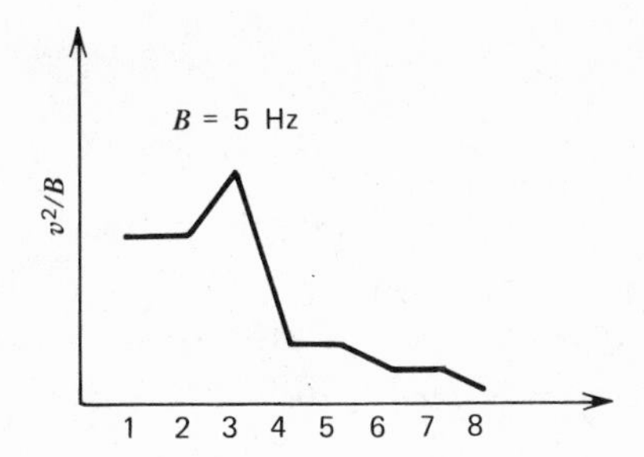

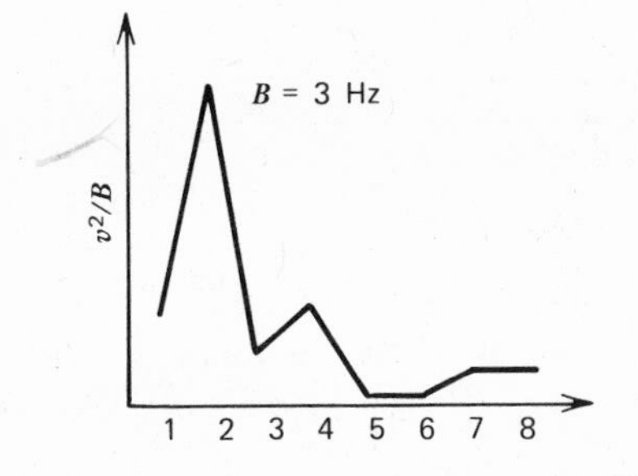

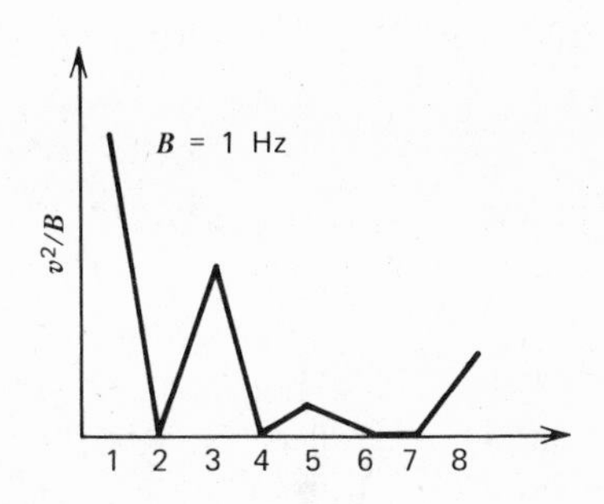

Figure 22

FACTORS TO CONSIDER IN SELECTION OF THE BANDPASS FILTER

1. When utilizing analog systems, the larger the B, the faster a plot can be generated.
2. The larger the B, the smaller the original data sample required. Sometimes long data samples are not available.

3. Usually large B's produce cost savings in analog and digital processing time.
4. The larger the B, the less distinctive actual modes appear (see Example 2).

Selection of a bandpass filter is a "give-and-take" proposition. Processing costs go down with increases in B, but so does final resolution. Depending on the time of the data sample available, resolution required, and turnaround time requirements, B can be selected. Spectral density analysis is generally performed by one of three methods:

1. Analog tracking filters.
2. Hybrid analyzers.
3. Digital computer.

The tracking filter is especially effective when small quantities of data are processed. Accuracy is outstanding. The hybrid analyzer employs an analog filter but works rapidly by compressing data before filtering. Computers, although expensive to use, become highly cost effective when large numbers of plots are generated.

The sections that follow discuss analog and hybrid systems presently in use and a fourth system, the *all-digital analyzer.*

THE TRACKING FILTER ANALYSIS SYSTEM

One method of performing power spectral or amplitude spectral density analysis (PSD or ASD) is by use of a data tape loop (containing the analog signal to be analyzed), a sweep oscillator, a tracking filter, and an X-Y plotter. These devices are discussed at length in the paragraphs that follow. In the basic spectral density analysis system (Figure 23) the tracking filter contains the bandpass filter and the sweep oscillator advances the x axis of the plotter while simultaneously "tuning" the tracking filter. One question that might come to mind is, how many seconds of data should the tape loop contain? The answer lies in the equation

$$T = \frac{n}{2B}$$

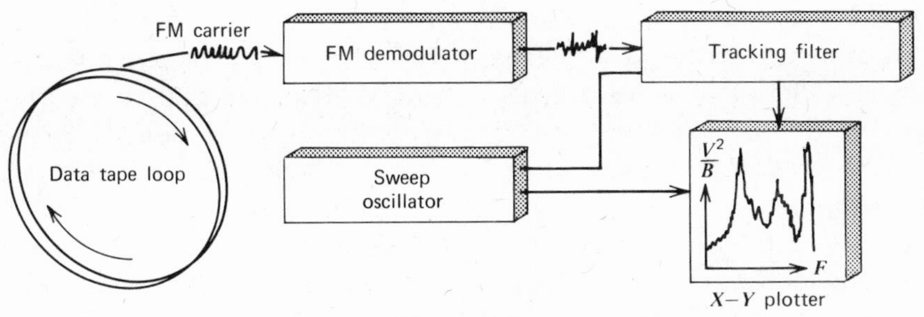

Figure 23

T is the minimum number of data seconds permissible, B is the value of the bandpass filter, and n is the number of degrees of freedom.

Degrees of freedom relate to statistical accuracy desired. A low n, for example, would occur in a straw poll if only New York City voters were questioned about the way they would vote in a state governor election. If city, rural, and suburban voters throughout the state were polled, n would be much higher. The higher the number of degrees of freedom, the more statistically accurate the final result.

Associated with each power spectral density plot is a number of degrees of freedom. When n is about 100, an accurate plot results; where n is low (e.g., about 30), plot accuracy is also low.

Actual PSD plot accuracy, with respect to degrees of freedom, may be derived from Figure 24. For about 33 degrees of freedom there is

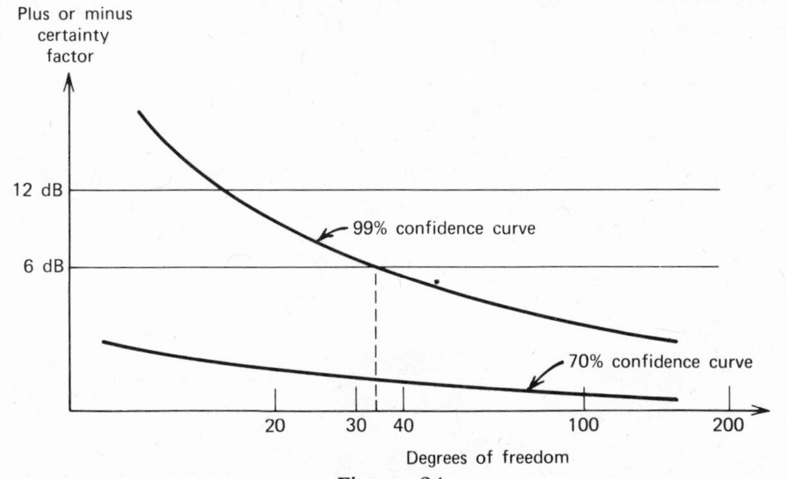

Figure 24

a 99% "certainty" that the *X-Y* plotted PSD curve will fall within ±6 dB of its true value. Also there is a 70% "certainty" that the final *X-Y* curve will rest in about ±2 dB of true value. Note that the figures quoted do not take into account the possible errors introduced by the actual plotting device. Additional errors of about ±2% are common.

It should be obvious from the graph that the higher the number of degrees of freedom the more accurate the final plot. Bearing in mind the equation $T = n/2B$, let us suppose that a data analyst required an accurate PSD plot (i.e., $n = 100$) of a 2.5-sec data sample. Assume also that he wished to use a bandpass filter of value $B = 10$ Hz. Therefore minimum sample time length $= 100/2 \times 10 = 5$ sec., but only 2.5 sec of data exist! In this case the analyst would be faced with one of two choices:

1. Change B to 20 Hz so that sample length $= n/2B = 100/2 \times 20 = 2.5$; that is, the exact number of data seconds available.
2. Leave B at 10 Hz and face the prospect of having a less accurate plot and one with poorer resolution.

Data signals requiring PSD processing usually contain frequency components ranging from f_L to f_U, where f_L is the lowest expected data frequency of interest and f_U is the highest.

The *sweep oscillator,* a device that causes analysis to begin, progresses through the frequency range of interest (the x axis); that is, f_L to f_U. As the oscillator's output frequency increases, the plotter's arm progresses to the right toward f_U.

As sweeping occurs, B (the tracking filter's bandpass filter) detects the amount of frequency component strength and routes this information to the plotter's y or v^2/B axis. More specifically, the sweep oscillator provides a "center frequency" to B; for example, if $B = 5$ Hz and the sweep oscillator is at value (x axis) 8 Hz, we have Figure 25. Note that the distance from 5.5 to 10.5 Hz $= 5$ Hz $= B$, and when

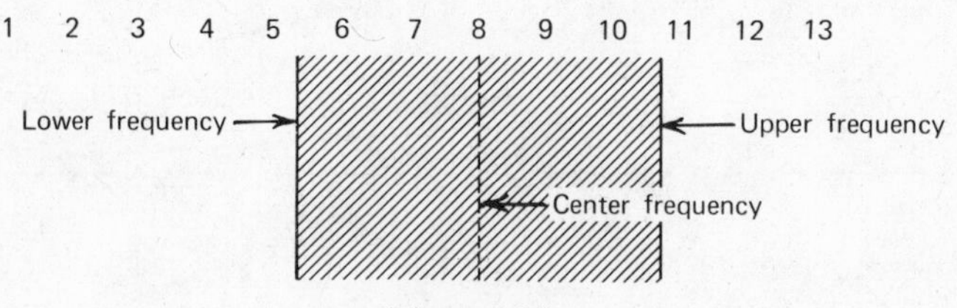

Figure 25

the sweep oscillator is at 8 Hz the tracking filter accepts components with frequency between 5.5 and 10.5 Hz.

As another example, if $B = 2$ Hz and the sweep oscillator were at 3 Hz, then frequency components in the 2 to 4 Hz range would be acted on.

It is important to note that when reading a PSD plot the v^2/B reading at any frequency point is one for those frequency components in the frequency band around and including that point. (Many have been mistaken by believing that the v^2/B reading pertains only to the x-axis frequency reading.)

The *tracking filter* is a frequency-tuned bandpass filter that accepts two inputs—a tuning signal f_T and the electrical signal to be analyzed; f_T is generated by the sweep oscillator and the electrical signal contains frequency components f_n, where f_n represents the summation of all frequency components in the original data signal.

A block diagram of the tracking filter and its principles of operation are shown in Figure 26.

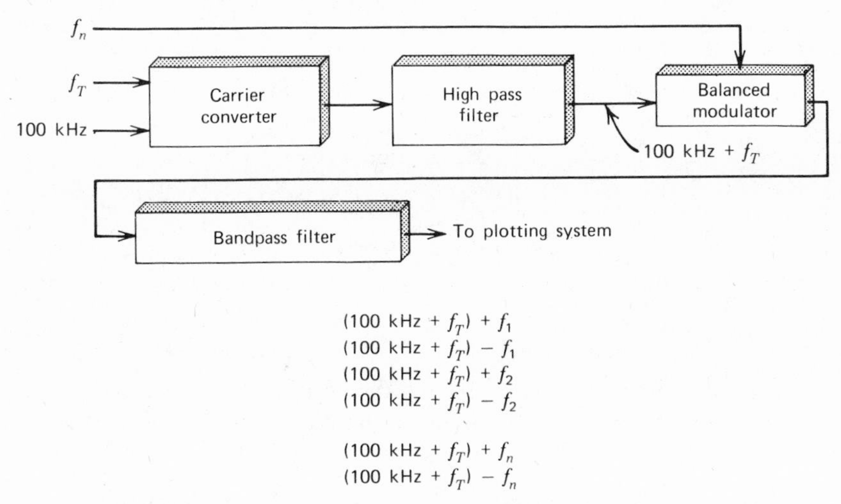

Figure 26. Block diagram, the analog tracking filter.

The tuning frequency, generated by the sweep oscillator, is mixed with a reference frequency; in this case, with 100 kHz. f_n represents all of the frequency components in the original data signal.

Out of the balanced modulator comes the sum-and-difference frequencies (of data components) and the carrier converter frequency.

These outputs are routed to the bandpass filter B whose output is an electric signal that represents the strength of data components, if any, at the tuning frequency.

In terms of an actual example, assume a vibratory data signal (e.g., from an accelerometer) containing four frequency components, each of which is at a different strength in terms of amplitude. The four frequencies are 5, 30, 80, and 265 Hz. Assume, too, that during the plotting operation we employ $B = 5$ Hz and observe what happens when the sweep oscillator (and x axis) is at 80 Hz.

- Into the converter go 80 Hz and the reference 100 kHz.
- Out of the converter comes 100,080 Hz.
- Routed to the balanced modulator are the 100,080-Hz signal, plus the data frequencies; that is, 5, 30, 80, and 265 Hz.
- The sum and difference frequency components appear at the output of the balanced modulator:

$$\begin{aligned}
100{,}080 + 5 &= 100{,}085\\
100{,}080 - 5 &= 100{,}075\\
100{,}080 + 30 &= 100{,}110\\
100{,}080 - 30 &= 100{,}050\\
100{,}080 + 80 &= 100{,}160\\
100{,}080 - 80 &= 100{,}000\\
100{,}080 + 265 &= 100{,}345\\
100{,}080 - 265 &= 99{,}815
\end{aligned}$$

- These sum-and-difference frequency components are routed to the bandpass filter whose center frequency is 100,000 and whose bandwidth is 5. This means that only data with frequency components in the 99,997.5 to 100,002.5 range can be bandpassed.
- Only 100,000 Hz passes through the filter. The amplitude of this signal is directly proportional to the magnitude of the original 80-Hz signal.

The tracking filter's 100,000-Hz output may be converted to a voltage and is suitable for routine to the X-Y plotter (at x axis = 80 Hz) or to an analog to digital converter if results are to be converted for computer entry.

As the frequency addition and subtraction operations take place, the changing, or sweeping up, of the tuning frequency must occur slowly enough so that the bandpass filter's response-time limitations do not cause inaccurate tracking of the variations in the amplitude of the 100-kHz signal. In plainer words, filters take time to work; for accurate results we must allow sufficient time at each x-axis setting. A recommended x-axis sweep rate for most tracking filter applications is

$$\text{sweep rate} = \frac{B}{2T}$$

T, the minimum sample length, may be derived from the equation discussed earlier, that is, $T = n/2B$.

As an example, suppose we require that a PSD plot be generated with the x axis beginning at 10 and ending at 2000 Hz (a total sweep distance of 1990 Hz). Suppose, too, that we are using $B = 10$ Hz, that our plot must have 120 degrees of freedom, and that our original data were recorded at tape speed = 15 in./sec. Calculating, first, the minimum sample length required, we obtain

$$T = \frac{n}{2B} = \frac{120}{2 \times 10} = 6 \text{ sec}$$

Calculating the sweep rate, we obtain

$$SR = \frac{B}{2T} = \frac{10}{2 \times 6} = 0.8 \text{ Hz/sec}$$

Because time equals distance divided by rate and sweep distance = 1990 Hz, we obtain, for total plotting time,

$$\text{time} = \frac{1990}{0.8} = 2487.5 \text{ sec} = 41^{+} \text{ min}$$

Conclusion: PSD plotting by tracking filter is time consuming. Often a means of speeding analysis time is achieved by a technique commonly termed *speed jumping*. Referring to the preceding example, suppose we doubled tape playback speed from 15 in./sec (ips) to 30 ips. What occurs is that twice as many data pass the tape transport's reproduce heads each second. Also, all frequency components formerly ranging between 10 and 2000 Hz actually double in frequency, thus creating a new range of 20 to 4000 Hz (a total new frequency span of 3980 Hz).

Under these conditions, that is, with all frequencies doubled, we must double B in order to perform power spectral density analysis. Making new calculations for the speed-jumped case, we obtain

Tape speed = 30 ips

Doubled $B = 20$ Hz
T = minimum sample length = $n/2B$ = 3 sec
Total x-axis span = 3980 Hz
Sweep rate = $B/2T$ = 3.33 Hz/sec
Total analysis time = total span/sweep rate = 3980/3.33 = approx. 1200 sec or 20 min.

Obviously with this technique analysis time is reduced significantly, but even more time could be saved by speed jumping to tape speed = 60 ips and using $B = 40$ or to 120 ips and a 80-Hz filter. Certainly speed-jumping or time-compressing data is advantageous but requires that additional hardware be obtained to permit the increasing of tape-loop speed and the doubling of B. Sometimes, though, the acquisition of additional hardware or the use of existing additional systems for the sole purpose of speed jumping is not always practical or economical.

Another analysis system that does not use the sweep oscillator/tracking filter combination is the hybrid or real-time analyzer. Discussed in the following section, it utilizes a highly effective digital speed-jumping technique.

THE HYBRID POWER SPECTRAL DENSITY ANALYZER

Speed jumping, as mentioned earlier, is highly desirable when rapid processing is required. More specifically, speed jumping permits the use of an extremely large bandpass filter that operates on data rapidly. Hardware limitations, however, usually prohibit speed jumping by factors greater than 8 or 16 and have led to the development of a digital technique that permits speed-jump factors of 100, 1000, and more. Reviewing the definitions of analog-to-digital and digital-to-analog conversion (discussed in Chapter 9), we have the following:

- The analog-to-digital (A/D) converter samples continuous analog signals and converts them into a series of binary digits (bits) which are either 1's or 0's. Each group of bits, called a digital data word, represents the analog value of each sample (Figure 27).
- The digital-to-analog converter (DAC) does the reverse (Figure 28).

Now consider the hardware configuration shown in Figure 29. The system was first used in the mid-60's and is hybrid in nature because it

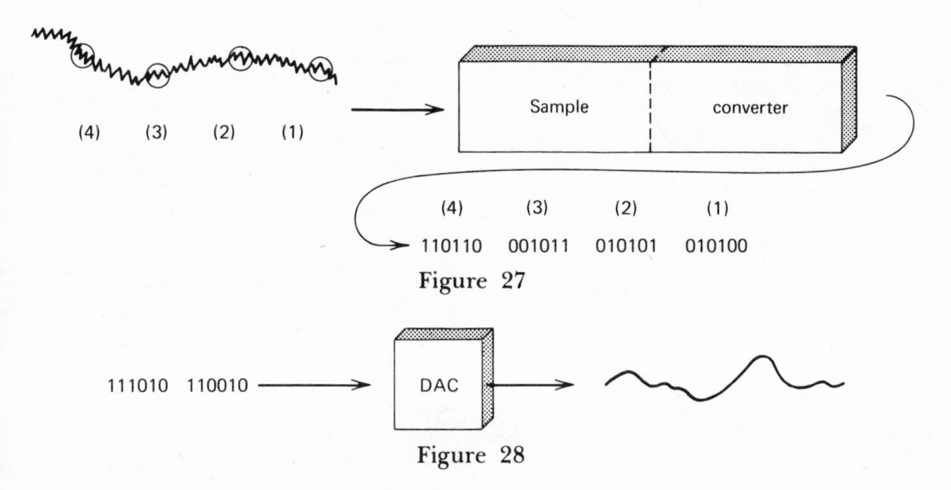

Figure 27

Figure 28

contains both analog and digital subsystems. Two primary functions are performed:

1. The first n data samples are converted to a parallel binary format and placed in the n memory locations in the circulating memory. (The memory system is capable of speed-jumping samples to factors of 2500 to 1.)
2. As soon as the next sample reaches memory, one of the older samples is discarded to make room for the new. Memories contain all samples in their order of arrival.

The circulating memory speed jumps original data and "spins" so fast that many old data words are recirculated in the time it takes for new data samples to be analog-to-digitally converted and made ready to "kick out" old samples.

As these two functions take place, the DAC "watches" memory contents and converts memory words into an analog signal. This signal, now highly time-compressed, is routed to a large bandpass filter for final processing.

By the use of high-speed digital subsystems, time compressions become so large that processing in the bandpass filter occurs almost instantaneously, or in *real time.* With tracking filters the rate of sweep has to be low enough to accommodate the filter's rise time. This problem is eliminated by time compression in the hybrid system. In any filtering operation the larger the filter, the faster the filtering operation.

Another point in the hybrid system is that if new data are not

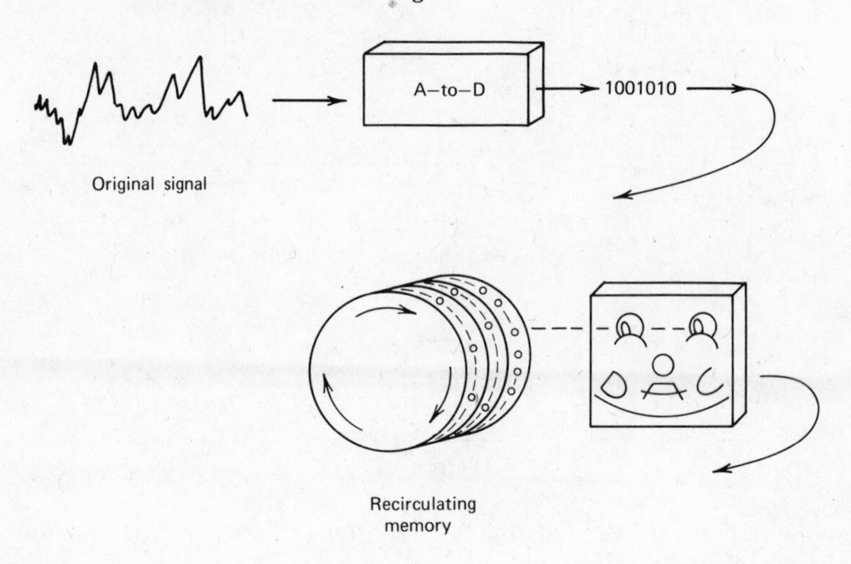

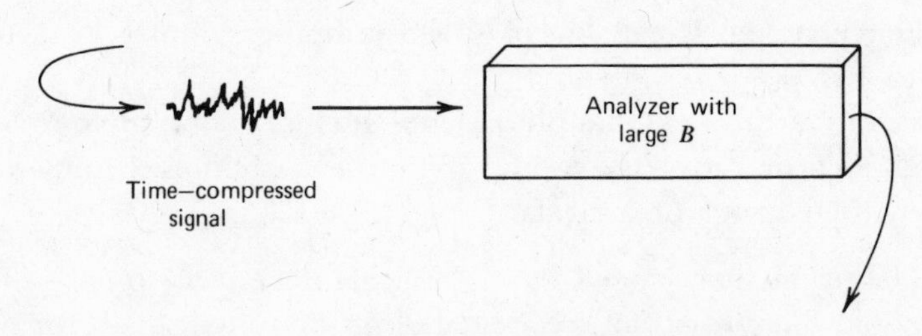

Figure 29

permitted to enter memory the resident signal will continue to circulate. This is especially significant, for if the exact location of an event such as a data anomaly or transient is known (e.g., from an oscillograph history containing IRIG B time code), the event alone can be fed into the analyzer for analysis.

SOME COMMENTS ABOUT THE ALL-DIGITAL ANALYZER

Although not yet widely used with respect to today's 2500- to 15,000-dollar analog filter analyzers, digital devices act on analog-to-digitally converted inputs in a Fourier or fast Fourier transform and can also perform autocorrelation, cross correlation, and several other operations. A problem, though, is that the maximum acceptable input frequency is about 50 kHz and all digital devices currently offer a lower dynamic range (i.e., about 60 dB versus 80 to 120 for devices containing analog filters). In time, however, such systems will improve in capability and will be in much use when more than just spectral density analysis is to be performed.

The all-digital analyzer, like the hybrid analyzer, can operate in the real-time mode. Both systems contain CRT's to display final plots.

SUMMARY—SPECTRAL DENSITY ANALYSIS SYSTEMS

The power spectral density or the amplitude spectral density plot may be generated by three commercially available systems:

1. The tracking filter/sweep oscillator/plotter system.
2. The hybrid, or real-time analyzer.
3. The digital computer or all-digital analyzer.

Certainly the cost of a computer and its associated software is high, but when vast quantities of data are to be processed the cost per plot by digital techniques goes down. When, for example, a large variety of data from a plane under flight test are telemetered to a digital computing facility, it is practical to process pressure, temperature, flow, and other parameters together with dynamic analysis data. When the computer is used, final displays (i.e., the PSD plots) may be routed to CRT and/or to a computer output microfilm system (COM).

No matter which system is chosen, final plots indicate the frequency composition of data (Figure 30).

Another dynamic analysis tool, the *autocorrelation function,* may also be used by analog or digital means. Analog considerations are discussed in the section that follows.

THE AUTOCORRELATION FUNCTION

The autocorrelation function $R(X)$ is a measure of the similarity between two waveform segments or, in other words, a function that

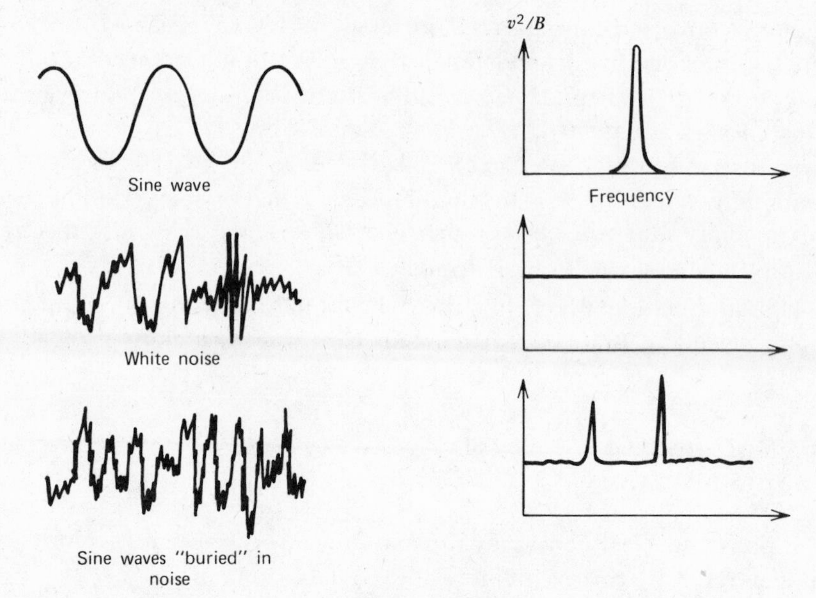

Figure 30

describes the dependence of data values at time t_1 on the values at another time $t_1 + X$.

How similar are two signal segments? In one method of measuring similarity they are multiplied together at each instant of time and their products are added over the duration of the signals. Once done, the similarity may be established by summing the products of time samples (see Figure 31). To evaluate the similarity between signal A and signal B we multiply a_1's value by b_1's, a_2's by b_2's, . . . , a_n's by b_n's and add their products.

$$(a_1 \times b_1) + (a_2 \times b_2) + (a_3 \times b_3) + (a_4 \times b_4) + \ldots (a_n \times b_n)$$
$$= (6.2 \times 6.2) + (5.0 \times 5.0) + (3.0 \times 3.0) + (3.0 \times 3.0) + \ldots (a_n \times b_n)$$
$$= 81.44 + \ldots (a_n \times b_n)$$

In mathematics a + times a + = a + and a − times a − also = a +. In the above case we will always be multiplying in this manner and the final product will be a large positive number.

Evaluating the similarity of A and C, we obtain:

$$(a_1 \times c_1) + (a_2 \times c_2) + (a_3 \times c_3) + (a_4 \times c_4) + \ldots + (a_n \times c_n)$$
$$= (+6.2 \times +3.9) + (+5.0 \times +1.2) + (+3.0 \times -2.5) + (+3.0 + -2.7)$$
$$+ \ldots + (a_n \times c_n) = (+24.18) + (+6.00) + (-7.50) + (-8.10)$$
$$+ \ldots + (a_n \times c_n) = +14.58 + \ldots + (a_n \times c_n)$$

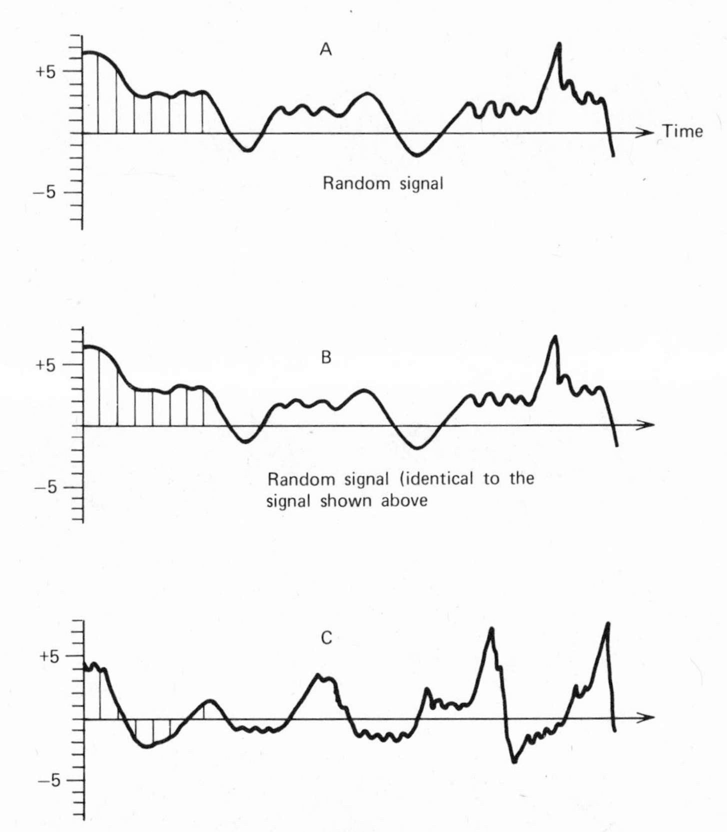

Figure 31

Here we have cases in which positive numbers were multiplied by negative numbers and + times − = a negative number. In this case the autocorrelation of *A* and *C* is considerably smaller than that of *A* and *B*. Consider the two waveforms in Figure 32. *A* and *B* are identical in shape but *B* is displaced in time. If we multiplied ordinates, we would find that many positive products would cancel many negative products. The more we time-shift signal *B*, the more number cancellations that as shown graphically in Figure 33. For random signals, if no time shift is created, the final sum is large. As the time shift is increased, the final sum will decrease, or the autocorrelation function for a random waveform decreases as the time shift increases.

The actual autocorrelation function is defined as follows: $R(X)$ of a waveform, $v(t)$, is a graph of the similarity of the waveform and the delayed (time displaced) replica $v(t + X)$ as a function of the time delay (X).

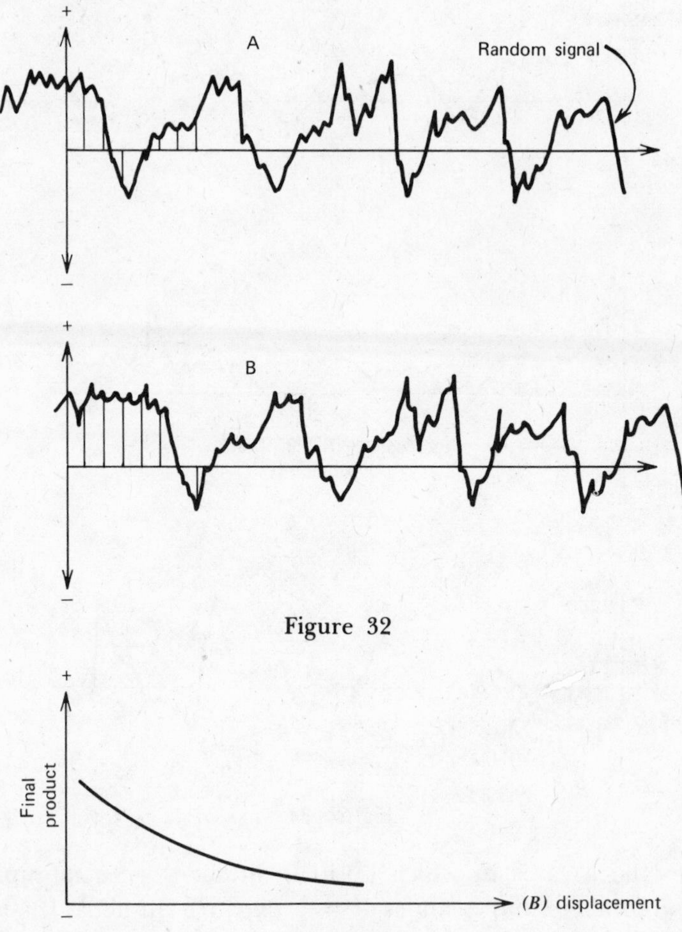

Figure 32

Final product

(B) displacement

Figure 33

$$R(X) = \frac{1}{T}\int_0^T v(t)\,(v(t + X))\,dt$$

where T = observation time,
$v(t)$ = waveform.

The autocorrelation of a sine wave is a cosine wave. A graphical explanation appears in Figure 34. At time displacement X_1, where $X_1 = 0$, the original signal and the displaced signal are in perfect synchronization (the dotted area). At this point $R(X)$ is maximum. As X increases to X_2, X_3, X_4, and X_5, the original and displaced signals

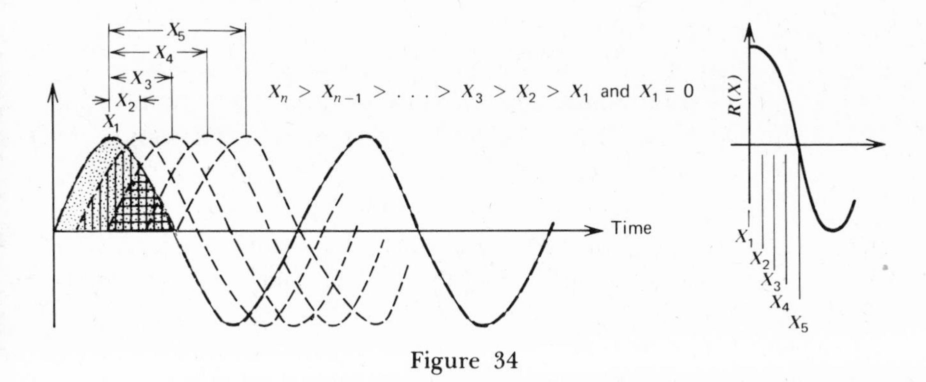

Figure 34

become less synchronized. If we continued this process through X_6, X_7, and so on, we would obtain the $R(X)$ cosine wave shown. $R(X)$ may be considered the amount of synchronization as a function of time displacement. Additional examples of autocorrelation plots are given in Figure 35.

NOTE. Wide-band random noise contains many frequency components, whereas narrow-band noise contains only low-frequency components.

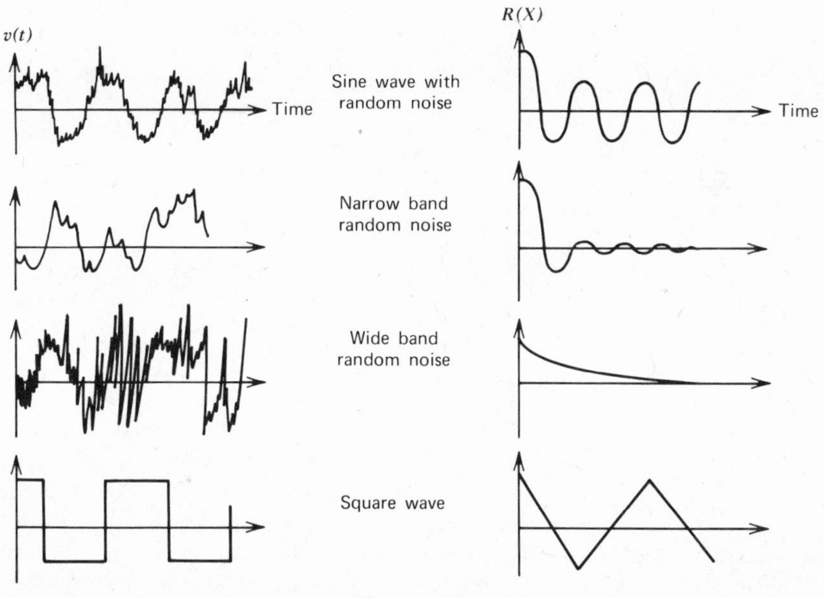

Figure 35

The autocorrelation function for each waveform is dependent on its frequency content. Periodic functions will yield periodic autocorrelation functions and nonperiodic signals yield a nonperiodic $R(X)$. Mixed functions, the sine wave with random noise, for example, will yield autocorrelation functions that are also mixed.

Each distinct waveform has a correspondingly unique autocorrelation result. Based on this fact, $R(X)$ plots are extremely useful in the detection of periodic data that might be hidden (masked) in a random background.

One way to perform autocorrelation plotting in the data processing laboratory is by use of a special tape transport that contains a storage bin. The number of inches of tape in the bin, while the system is running, is directly related to and determines the time displacement X. $R(X)$ is estimated by delaying the original signal $v(t)$ to create $v(t + X)$ and multiplying $v(t)$ values by the $v(t + X)$ values appearing on the tape transport's second read head. Multiplication products are averaged over the observation time (Figure 36.)

The autocorrelation function, which is highly dependent on the frequency content of a waveform acquired by the transducer or a

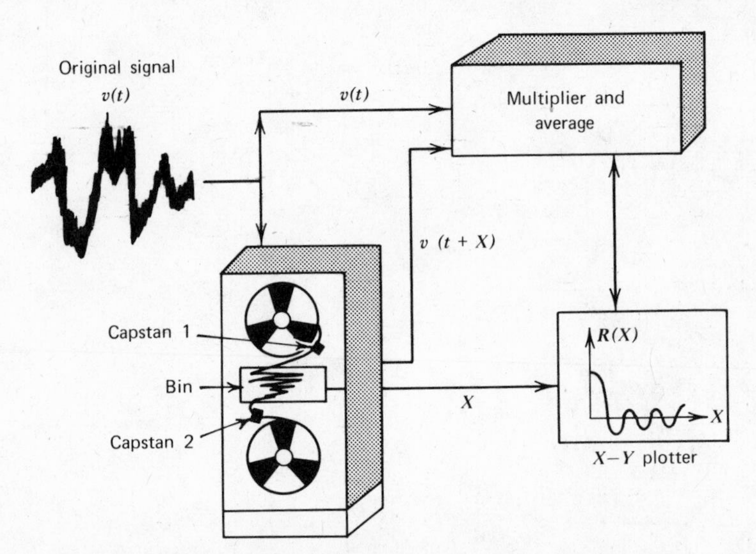

Figure 36. Two capstans that drive tape past read heads are contained in the system. If capstan 2 operates more slowly than capstan 1, tape buildup occurs in the bin. X, the time delay between data read at head 1 versus head 2, increases as capstan 2 decreases in speed.

remote sensor, may be used to extract meaningful periodic data from a data signal containing electrical noise. More precisely, deterministic data hidden in a random background (described earlier) may be extracted by the sophisticated system shown in Figure 36, or by other systems which include the digital computer that can act on a sensor's output if it is first analog-to-digitally converted.

The cross-correlation function $R_{vw}(X)$ is similar to the autocorrelation function. Processing methods are similar but the cross-correlation function is used when two nonidentical waveforms exist; essentially it is a measure of the similarity between different waveforms. We discuss it in the section immediately following.

CROSS CORRELATION

Cross correlation is a means of measuring the similarity between two nonidentical waveforms and is represented by an $X - Y$ plot. Cross-correlation plots are extremely useful in the following ways:

- Detecting signals buried (masked) in noise.
- Establishing the frequency response of a system.
- Determining a target's distance.
- Determination of the direction from which acoustic or electrical disturbances eminate.
- Establishing the transmission paths of electrical or vibratory signals.
- Decoding of digital or binary signals.
- Measurement of time delays caused by a system.

A cross-correlation estimate is obtained in the same manner as the autocorrelation function.

1. By delaying one signal $v(t)$ with respect to another signal by X seconds.
2. By multiplying $w(t)$ by the value of $v(t)$ that occurred X sec before.
3. By averaging multiplication products over the sampling time T.

Mathematically, the cross-correlation function is

$$R_{vw}(X) = \underset{T \to \infty}{\text{limit}} \frac{1}{T} \int_0^T v(t)\, w(t + X)\, dt$$

A typical cross-correlation plot for a pair of random signals ($v(t)$ and $w(t)$) appears in Figure 37. Note that the plot indicates peaks that

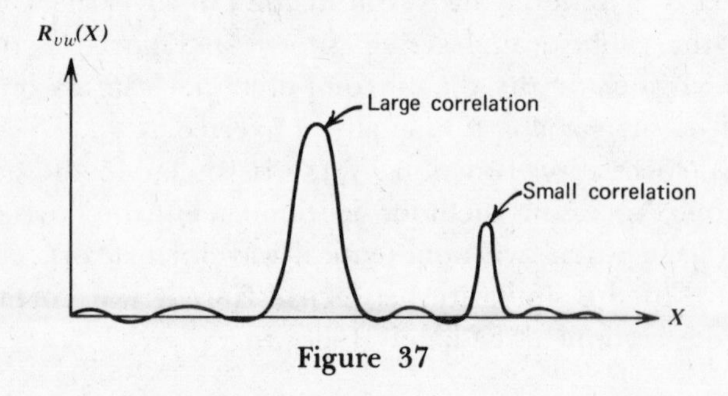

Figure 37

show definite correlation between $v(t)$ and $w(t)$ for specific time delays (X's) between the signals. The equipment used to effect the cross-spectral density analysis is the same as that used to generate the autocorrelation plot (Figure 38).

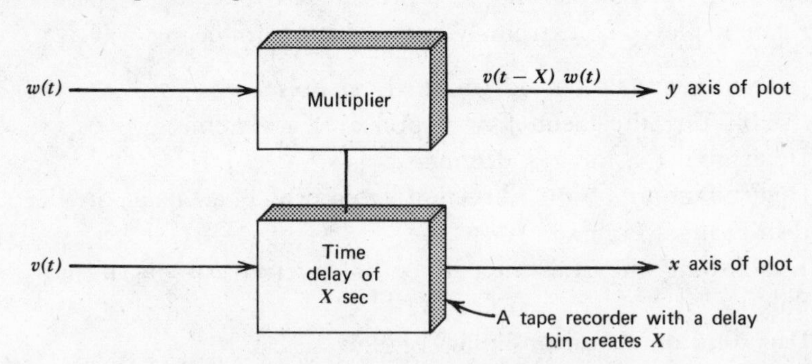

Figure 38. The cross-correlation system.

Among the numerous scientific and engineering applications for $R_{vw}(X)$ is finding a target's distance by correlating a transmitted sonar or radar signal with the signal reflected from a target (Figure 39).

When the transmitted signal is delayed by $(A + B)$, the transmitted and received signals will contain similar waveforms, thus causing the cross correlation to be large (Figure 40). When the large $R_{vw}(X)$ occurs, the reflected signal has returned. Knowing the time of return (X) and the speed of propagation through the medium we are concerned with (e.g., water, air, and rock), we may then calculate the target distance with the equation distance = rate × time. (Time is obtained from the x axis of the cross-correlation plot.)

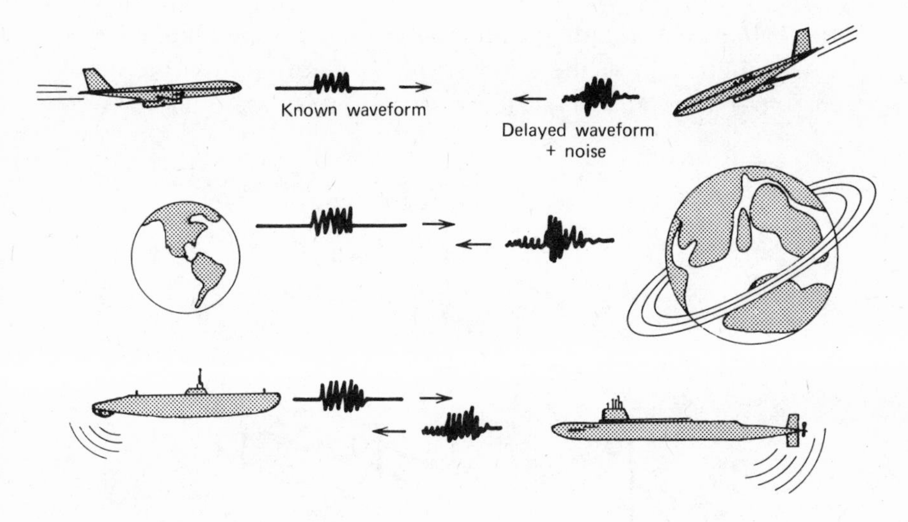

Figure 39. A is the time for the signal to reach the target; B is the time for the signal to return.

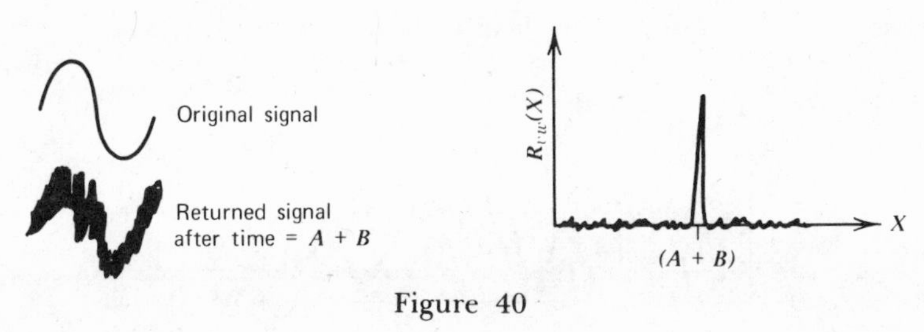

Figure 40

The final $X - Y$ plot may also be used to determine the reflectivity of the target. The actual reflectivity factor may be established by the magnitude of the cross-correlation function and a knowledge of propagation losses through the subject medium.

In addition to the cross correlation, the autocorrelation, and the power spectral density functions, there is a fourth—the probability density function. It, too, is useful when the physical properties of data must be established. Its definition is given in the section that follows.

PROBABILITY ANALYSIS

Consider, as we did in discussing stationary data, the effects of the wind on a flag. Assume in this case that the flag is located in the tropics, where never-ending and random trade winds blow. Let $h(t)$ represent the height of the lower corner of the flag from the ground as time goes on (Figure 41). Figure 42 represents a plot of $h(t)$ versus

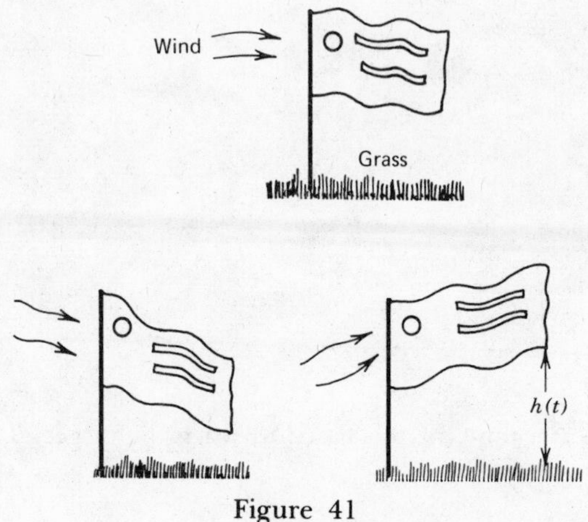

Figure 41

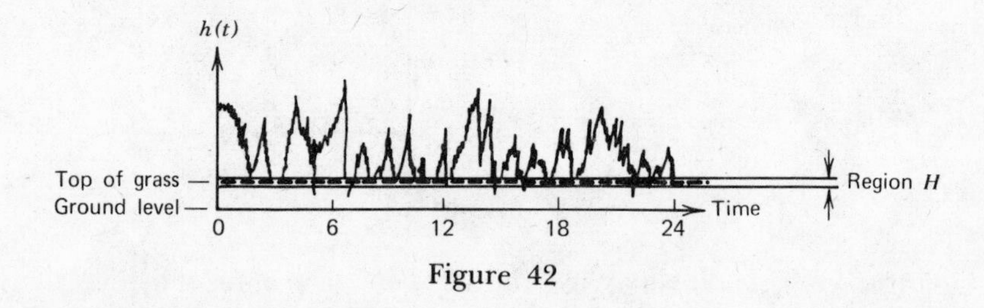

Figure 42

time for one day. The *probability density function* is defined as T_h/T_t, where T_h is the time $h(t)$ falls within a defined height zone H and where T_t is the total observation time. It should be noted that the probability density function is always a number ranging in value between 0 and 1; 1 indicates a 100% probability.

If T_h/T_t were 0.035 for region H (i.e., near the grass tops), it would mean a 3.5% probability that at any given time of day the bottom of the flag would rest in the grass-tops region.

In order to find out what the probability is that $h(t)$ would be less than or equal to some specific value of $h(t)$, we turn to the *probability distribution function* which is an integral of the probability density function from minus infinity to $h(t)$.

The Probability Density Function of a Sine Wave

The final function (shown as an $X - Y$ plot) describes the probability that test data from the transducer or remote sensor will fall within defined y-axis ranges during each instant of time (Figure 43). Each unique waveform, when analyzed, yields a unique probability plot. Two such plots are shown in Figure 44. Given any transducer or sensor waveform, the probability density plot can be generated. The

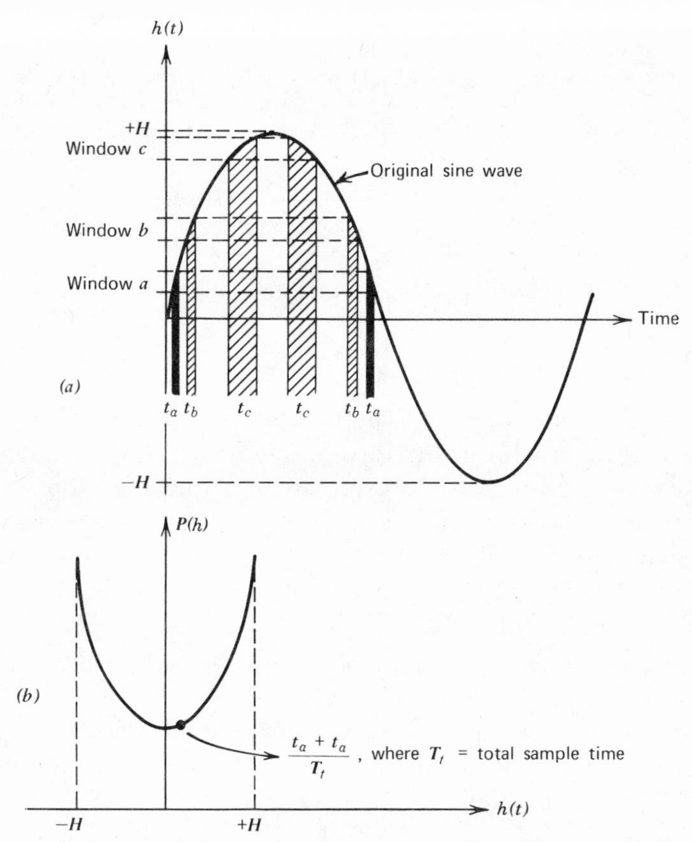

Figure 43. Note that as the amplitude window moves up the y axis (i.e., from a to b to c) the amount of time that $h(t)$ spends in the window increases. Thus the probability density value increases as we approach curve peaks from the origin. The probability density plot, for one complete sine wave cycle is shown in (b). Note also that probability density P(h) is lowest at the origin but highest from peak areas around + and − H.

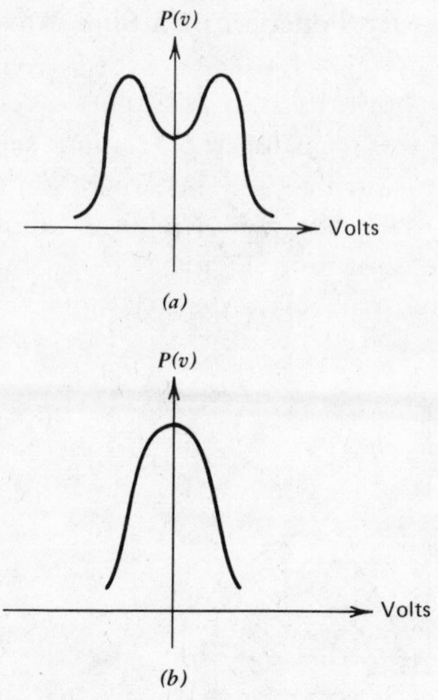

Figure 44. (*a*) Probability plot from a sine wave and random noise; (*b*) probability plot from random data.

general shape of the curve usually falls into one of several categories. Thus curves are useful in the identification of a "mystery" signal.

In the analog probability density analyzer a voltage window is swept between selected lower and upper values. The voltage window, in effect, performs a bandpass operation per amplitude band. Detected time that signals spend in the band is divided by T_t; that is, the total time of the data sample, and the result is routed to the plotter's y axis. As the voltage window is swept from value h_1 to h_n, so is the plotter's x axis.

In summary, the analog system (an analyzer and plotter) amplitude filters measure time spent in each amplitude region of interest, perform the probability density calculation, and route the results to an $X - Y$ plotter.

CROSS-SPECTRAL DENSITY ANALYSIS

Still another highly valuable tool to establish the physical characteristics of data is the cross-spectral density function (CSD). It serves to measure the relations between two voltage or time history records; for example, it can be used to establish the relation between a signal entering a system and a signal leaving a system (Figure 45). The CSD

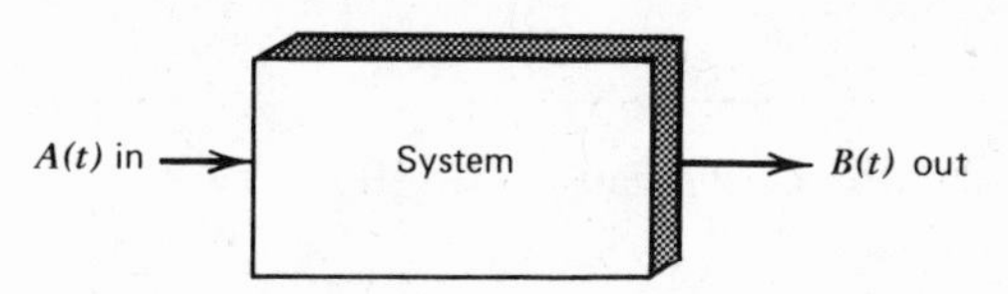

Figure 45. $A(t)$ and $B(t)$ represent electrical or mechanical measurements taken over some time period.

function consists of two parts, the cospectral density function and the quadriture spectral density function. The first is the averaged product of $A(t)$ and $B(t)$ in a bandpassed frequency interval, divided by that interval. The second is the same as the cospectral function except that either $A(t)$ or $B(t)$, not both, is shifted in phase by 90°. Mathematically,

$$\text{CSD} = \frac{1}{BT}\int_0^T A(t)\,B(t)\,dt - j\left(\frac{1}{BT}\int_0^T A(t)\,B'(t)\,dt\right)$$

B′(t) incorporates the 90° phase shift.

Output data, or the plots from a CSD processing system, reveal information regarding energy stored and/or phase shifted by the system under study (Figure 46). CSD plots may be generated by the analog system as follows:

1. $A(t)$ and $B(t)$ are filtered by identical bandpass filters.
2. The x axis of the final plot follows the upsweep of the center of the bandpass filters as it moves through the desired frequency range.
3. The two filtered signals are multiplied.
4. $A(t)$ filtered is multiplied by a 90° phase-shifted $B(t)$ that has also been filtered.
5. The multiplication products are averaged over time.
6. The averaged results are divided by the frequency bandpass filter's value B.

During analysis, as the center frequency of B is changed, a plot of real and imaginary components is generated.

The actual CSD system can be constructed with two power spectral density analysis systems (e.g., tracking filter systems), plus multipliers, and a device containing 90° phase shifting circuitry. System output

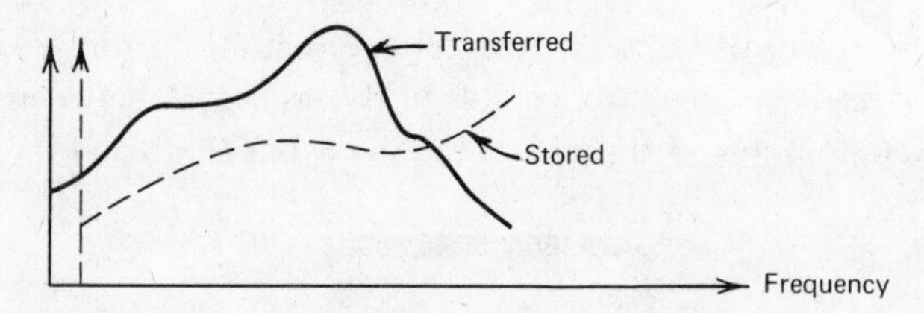

Figure 46. Optimum energy transfer is effected when the transferred curve is maximum.

plots are useful in several ways. These plots include, but are not limited to, the following:

- Determination of the amount of energy transferred into and out of a system.
- Determination of phase shift and time delay through a system.
- Establishment of a system's frequency response.

Usually, when a physical system needs describing, the CSD, the PSD, the autocorrelation, the probability density, or one of several other dynamic analysis functions is highly useful. For each type of analysis both analog systems and specialized digital computer programs capable of yielding the required plots are available.

13

EARTH-RESOURCES DATA HANDLING

Throughout the world today there are growing populations and their associated effects on the environment, Thus there is a pressing need to map and monitor natural and cultural resources periodically in an accurate and cost-effective manner. In many areas changes occur at rates with which map makers now find it difficult or impossible to keep pace.

However, by the use of specially tailored *in situ* water or ground data, as well as data from air- and spacecraft, man is able to perform studies related to agriculture, geology, water resources, environmental quality, and land use. Proper handling of multiband camera, *in situ* transducer, and remote sensor data permits environmental surveys of vegetation cover, water turbidity, strip mining, sea ice, urban sprawl, coral reefs, and a host of other earth features.

The success of earth-resources survey and mapping programs is often based primarily on their ability to extract meaningful data properly from the vast quantities that are made available from data acquisition systems. Satellite-to-ground data rates of 15 megabits, for example, necessitate data processing procedures tailored to extract and operate only on those data representing geographical regions of interest, anomalies, or feature changes with respect to time. In the 185 × 185 km scene data made available by the Earth Resources

Technology Satellite (ERTS)* a digital computer printout of multispectral scanner data for four spectral bands (MSS bands 4, 5, 6, and 7) measures about 1 ft in paper height! Obviously special computer processing techniques are required to summarize results or to display only specific geographic or anomalous areas. Figure 1 is an example of a "first look" computer-generated graphic summary of approximately 45,000 satellite pixels (picture elements) or, more specifically, scanner readings that were obtained along Puerto Rico's north coast by ERTS-1. Each pixel represents energy detected in an area of about 57 × 79m.

NOTE. A "second look" contour plot of this same Puerto Rico region is illustrated in Chapter 3. Its contour lines were generated in a computer program from a computer compatible tape. MSS band 7 data were extracted and are most useful in establishing land/water boundaries (Figure 1).

A further example of the need for specialized data-handling procedures can be demonstrated by considering data from a characteristics study of ocean currents.

If, for example, four test sites were to be studied by *in situ* current meters and current speed and direction were recorded on magnetic data tapes or microfilm once every 15 minutes, then, assuming a four-month data acquisition program, total data to be processed are

$$(4 \text{ sites}) \times (120 \text{ days}) \times (96 \text{ readings/day}) \times (2 \text{ parameters/reading}) = 92{,}160 \text{ points}$$

Following 16-mm microfilm-to-computer card conversion, or ¼in. analog FM tape processing (depending on the type of ocean current meter in use), we could call on the computer to list all recorded measurements chronologically. Here, however, final listings would require the data analyst to handle several hundred computer output line-printer pages.

As mentioned earlier, the key to a successful earth-resources program is the ability to extract meaningful data from the huge quantities of data often required. In the above case the data analyst might have been served best by a graphical summary of water current activ-

*ERTS was renamed Landsat in early 1975 following the launch (January 1975) of Landsat-2.

Figure 1. Contour plot generated from satellite data.

ity or listings generated only when current speed or direction fell outside expected or predefined ranges. Such graphs and listings are easily generated by computer, either to the line printer, to CRT, to an electrostatic printer plotter, or to graph-producing pen-to-paper plotters.

Earth-resources programs generally require one or more of the following data types:

1. Aircraft black and white and color photography, including stereo pairs.
2. Air- or spacecraft multiband photography.
3. Air- or spacecraft multispectral scanner and thermal IR scanner imagery and tapes. (Tape data are usually analog FM or digital PCM.)
4. Historical data, including maps, logs, and listings.
5. Ground truth, including but not limited to

- boat logs,
- laboratory reports,
- field data logs,
- ocean meter strip charts, magnetic data tapes, or microfilm,
- ground data collection system results (paper, film, or tape),
- weather logs,
- data collection platform (DCP) results.

6. Radar findings.
7. Magnetometer tapes.

In the design of an earth-resources data acquisition program we must first become acquainted with the areas to be surveyed so that appropriate cameras, sensors, and *in situ* transducers can be employed. Also, during the program design phase, it is a recommended procedure that data acquisition, data processing, and data analysis personnel discuss and establish criteria pertaining to initial data-handling requirements and potential retrospective data needs.

Often in the comprehensive data acquisition, processing, analysis, and application program data processing plays the chief role. In many projects 50% or more of the dollarcost goes for data handling. Thus the importance of data planning meetings, held well in advance of data acquisition periods, must not be underestimated; they permit the data processing facility to configure hardware to match the specifications of acquired data and enable the analyst to obtain results specifically formulated to his requirements. In addition, and perhaps the most relevant factor, proper planning prevents inundation of the analyst with superfluous or difficult-to-interpret listings, graphs, and imagery.

Earth-resources data handling is perhaps best summarized by a flow diagram (Figure 2). Obviously the data analyst is not restricted to listings of what happened where and when. By the use of specialized analog and digital techniques both graphic displays and map overlays (in black and white or color) can play the major roles. Design considerations for the environmental data base are discussed in Chapter 15. Also discussed are the digitizer—a valuable tool in establishing base maps onto which survey data may be presented,—and recommended display methods for aircraft, satellite, and ground (or water) data.

COMPUTERIZED RESOURCE INVENTORY SYSTEMS

Whenever large quantities of map or map-relatable data are to be acquired, a computer inventory system can be established. Some systems, like those in New York and Puerto Rico, for example, are based on grid overlays to maps and photos. Grid cells that represent an area of $1 km^2$ or a comparable measurement can be assigned an identity number (e.g., the geographical coordinates of each cell's lower left corner). Since 100 ha = $1 km^2$, map data can be related to hectares or percentages of kilometer cells.

Map or photo data, appearing through the transparent grid, are usually coded with symbols that reflect land-use categories, judged appropriate by the photo interpreter or the map reader. Coding for computer entry can be accomplished by counting the number of hectare cells per classification that exist in each grid cell. Results are entered into the computer by keypunched cards or a computer terminal.

Point data findings, such as miles of a feature (e.g., a highway), the number of features per cell (e.g., farm fields), or even a code to indicate the presence or absence of a dominant geologic, soil, or other type of land feature can be added for each cell. Once these coded

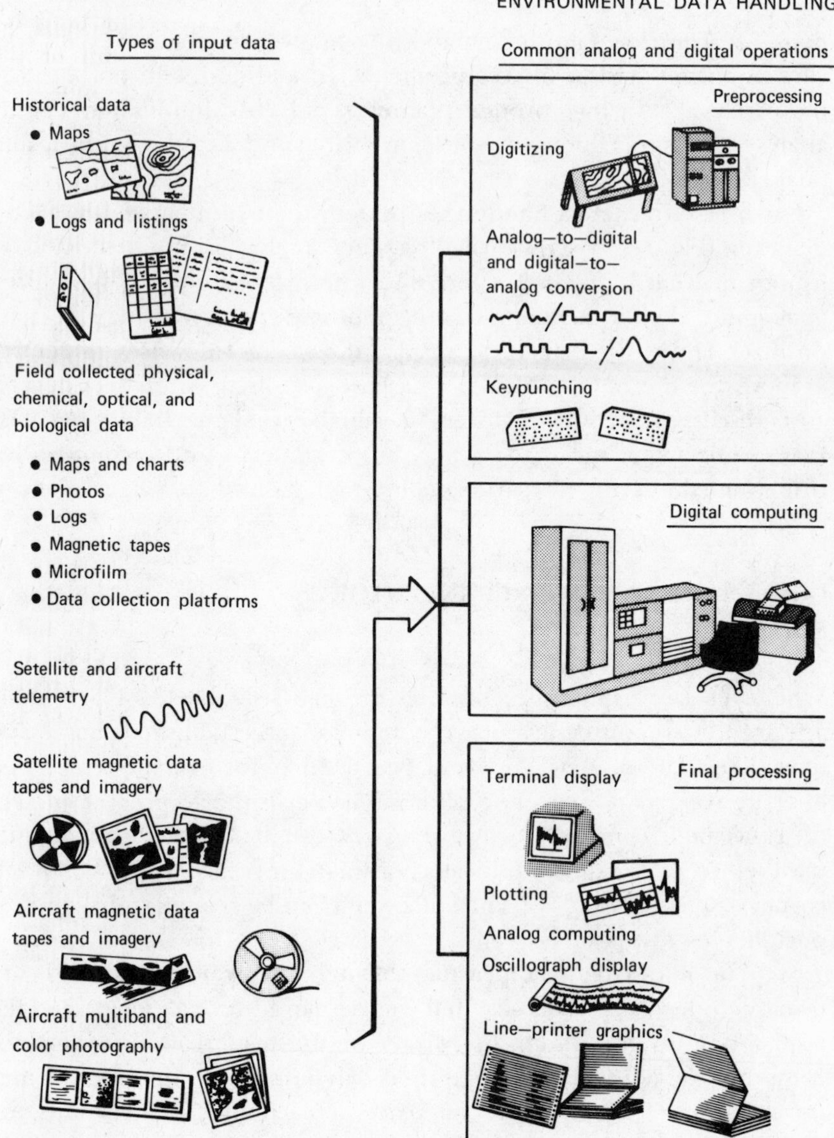

Figure 2. Environmental data handling.

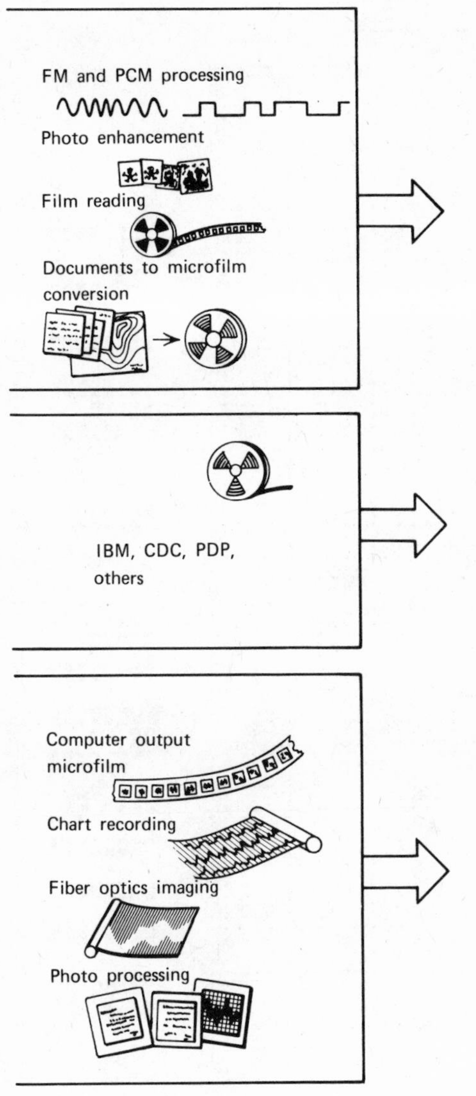

Common output data formats

Computer maps
- Overlays to standard maps
- Gray scale shaded maps
- Perspective plots

Statistical analyses
- Degrees of correlation
- Factor analysis tables
- Spectral density plots
- Probability plots

Pattern recognition
- Cluster plots
- Spectral signature maps

Contour plots

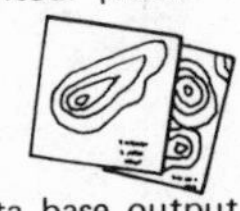

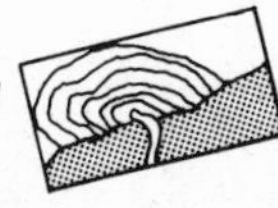

Data base output
- Measurement listings, graphs
- Status updates
- Measurement histories, trends

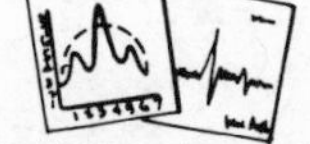

Imagery
- Paper or film
- Black and white or color

Simulations, models

Computer output microfilm

Cathode ray tube displays
- Black and white or color

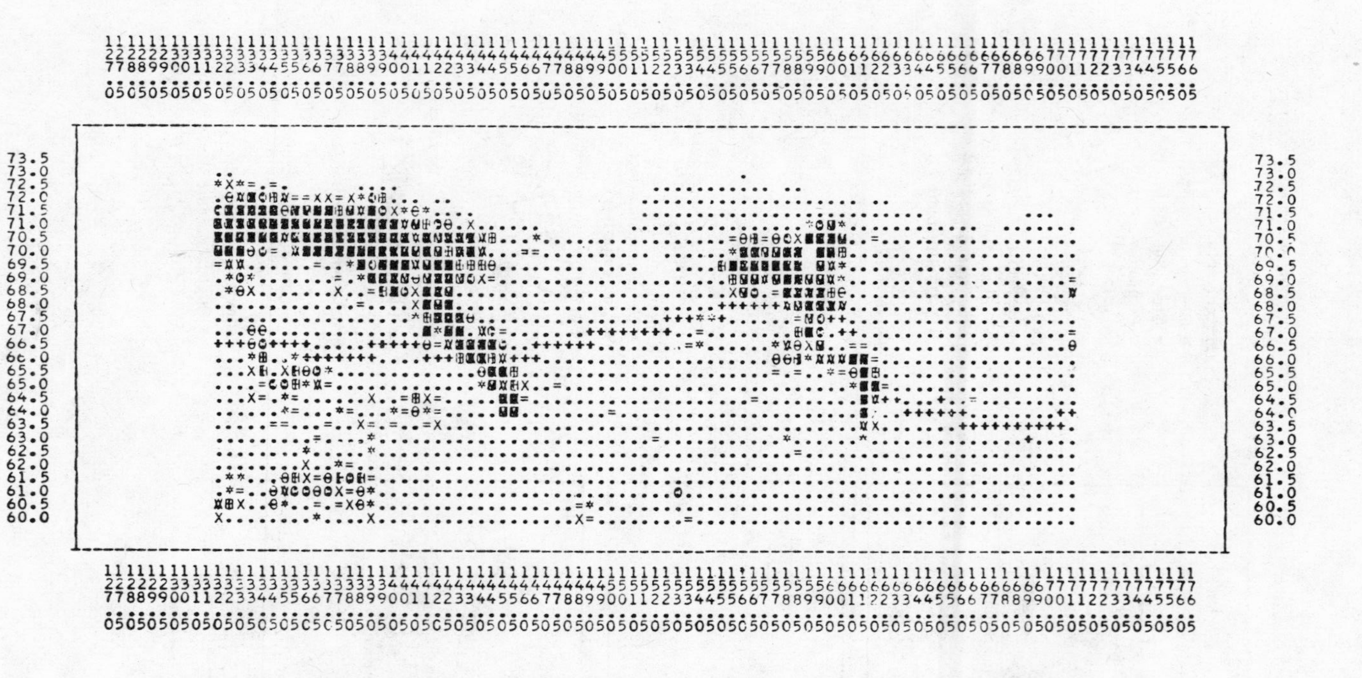

Figure 3. Actual computer inventory output map. Gray scale shades relate to percent of sugar cane per cell. Plus signs (+) indicate the locations of major highways. The region shown is the north central coastal zone of Puerto Rico. Blank cells are part of the Atlantic Ocean. Coordinate listings (latitude and longitude) also appear.

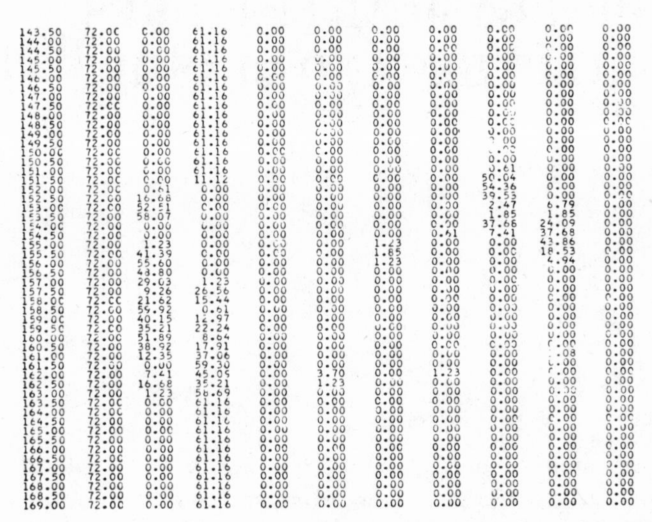

LAND USE TOTALS IN ACRES

EAST	NORTH	AGRIG	WATER	INDUS	FORES	COMM	RECRE	PUB.	RESID	SHYHY
143.50	72.00	0.00	61.16	0.00	0.00	0.00	0.00	0.00	0.00	0.00
144.00	72.00	0.00	61.16	0.00	0.00	0.00	0.00	0.00	0.00	0.00
144.50	72.00	0.00	61.16	0.00	0.00	0.00	0.00	0.00	0.00	0.00
145.00	72.00	0.00	61.16	0.00	0.00	0.00	0.00	0.00	0.00	0.00
145.50	72.00	0.00	61.16	0.00	0.00	0.00	0.00	0.00	0.00	0.00
146.00	72.00	0.00	61.16	0.00	0.00	0.00	0.00	0.00	0.00	0.00
146.50	72.00	0.00	61.16	0.00	0.00	0.00	0.00	0.00	0.00	0.00
147.00	72.00	0.00	61.16	0.00	0.00	0.00	0.00	0.00	0.00	0.00
147.50	72.00	0.00	61.16	0.00	0.00	0.00	0.00	0.00	0.00	0.00
148.00	72.00	0.00	61.16	0.00	0.00	0.00	0.00	0.00	0.00	0.00
148.50	72.00	0.00	61.16	0.00	0.00	0.00	0.00	0.00	0.00	0.00
149.00	72.00	0.00	61.16	0.00	0.00	0.00	0.00	0.00	0.00	0.00
149.50	72.00	0.00	61.16	0.00	0.00	0.00	0.00	0.00	0.00	0.00
150.00	72.00	0.00	61.16	0.00	0.00	0.00	0.00	0.00	0.00	0.00
150.50	72.00	0.00	61.16	0.00	0.00	0.00	0.00	0.00	0.00	0.00
151.00	72.00	0.00	61.16	0.00	0.00	0.00	0.00	0.61	0.00	0.00
151.50	72.00	0.00	11.12	0.00	0.00	0.00	0.00	50.04	0.00	0.00
152.00	72.00	0.61	0.00	0.00	0.00	0.00	0.00	54.36	0.00	0.00
152.50	72.00	16.68	0.00	0.00	0.00	0.00	0.00	39.53	0.00	0.00
153.00	72.00	52.51	0.00	0.00	0.00	0.00	0.00	2.47	6.79	0.00
153.50	72.00	58.07	0.00	0.00	0.00	0.00	0.00	1.85	1.85	0.00
154.00	72.00	0.00	0.00	0.00	0.00	0.00	0.00	37.66	24.09	0.00
154.50	72.00	0.00	0.00	0.00	0.00	0.00	0.61	7.41	37.68	0.00
155.00	72.00	1.23	0.00	0.00	0.00	1.23	0.00	0.00	43.86	0.00
155.50	72.00	41.39	0.00	0.00	0.00	1.85	0.00	0.00	18.53	0.00
156.00	72.00	55.60	0.00	0.00	0.00	1.23	0.00	0.00	4.94	0.00
156.50	72.00	43.80	0.00	0.00	0.00	0.00	0.00	0.00	0.00	0.00
157.00	72.00	29.03	1.23	0.00	0.00	0.00	0.00	0.00	0.00	0.00
157.50	72.00	9.26	26.56	0.00	0.00	0.00	0.00	0.00	0.00	0.00
158.00	72.00	21.62	15.44	0.00	0.00	0.00	0.00	0.00	0.00	0.00
158.50	72.00	59.92	0.61	0.00	0.00	0.00	0.00	0.00	0.00	0.00
159.00	72.00	40.15	12.97	0.00	0.00	0.00	0.00	0.00	0.00	0.00
159.50	72.00	35.21	22.24	0.00	0.00	0.00	0.00	0.00	0.00	0.00
160.00	72.00	51.89	8.64	0.00	0.00	0.00	0.00	0.00	0.00	0.00
160.50	72.00	38.92	17.91	0.00	0.00	0.00	0.00	0.00	0.00	0.00
161.00	72.00	12.35	37.06	0.00	0.00	0.00	0.00	0.00	[illegible].08	0.00
161.50	72.00	0.00	59.30	0.00	0.00	0.00	0.00	0.00	0.00	0.00
162.00	72.00	7.41	45.05	0.00	3.70	0.00	1.23	0.00	0.00	0.00
162.50	72.00	16.68	35.21	0.00	1.23	0.00	0.00	0.00	0.00	0.00
163.00	72.00	1.23	58.69	0.00	0.00	0.00	0.00	0.00	0.00	0.00
163.50	72.00	0.00	61.16	0.00	0.00	0.00	0.00	0.00	0.00	0.00
164.00	72.00	0.00	61.16	0.00	0.00	0.00	0.00	0.00	0.00	0.00
164.50	72.00	0.00	61.16	0.00	0.00	0.00	0.00	0.00	0.00	0.00
165.00	72.00	0.00	61.16	0.00	0.00	0.00	0.00	0.00	0.00	0.00
165.50	72.00	0.00	61.16	0.00	0.00	0.00	0.00	0.00	0.00	0.00
166.00	72.00	0.00	61.16	0.00	0.00	0.00	0.00	0.00	0.00	0.00
166.50	72.00	0.00	61.16	0.00	0.00	0.00	0.00	0.00	0.00	0.00
167.00	72.00	0.00	61.16	0.00	0.00	0.00	0.00	0.00	0.00	0.00
167.50	72.00	0.00	61.16	0.00	0.00	0.00	0.00	0.00	0.00	0.00
168.00	72.00	0.00	61.16	0.00	0.00	0.00	0.00	0.00	0.00	0.00
168.50	72.00	0.00	61.16	0.00	0.00	0.00	0.00	0.00	0.00	0.00
169.00	72.00	0.00	61.16	0.00	0.00	0.00	0.00	0.00	0.00	0.00

Figure 4. Computer inventory listing of dominant land-use features by acreage.

data have been presented to a computer data bank, the users of inventory data may, either at the computer or at a remote terminal (even many miles away), call for feature maps or listings of feature types per region. A feature summary map and an inventory listing follow. Note that inventory maps can also be produced in color.

Finally, the well-designed computerized resources-inventory system must be flexible enough to accept new data entries and modifications that are resident in data-bank files. (Figures 3 and 4).

14

THE DIGITAL COMPUTER FOR PHYSICAL-FEATURES DETECTION AND CLASSIFICATION

Aircraft or satellite spectral band data for a geographical area of interest may be subjected to computerized band ratio and ratio clustering analyses. Results often provide the data analyst with a basis for establishing that computer pattern recognition is possible within the scene of interest. When favorable, computer mapping programs may be applied.

Use of pattern-recognition techniques is by no means a simple matter. This is partly because detected radiation per band from a single type of ground feature rarely yields *identical* spectral characteristics for each land (or water) area sampled. Response-curve differences may occur in one or more variables, which in agricultural studies can include the following:

1. Crop-planting date.
2. Extent of fertilization, pesticides, and so on.
3. Soil moisture.
4. Atmospheric pollutants.

The extent of variation in one feature's response curves (e.g., fields of beans) must be compared with the response curves of surrounding features in order to determine whether the digital computer can classify features boundaries accurately. It is entirely possible that although beans and other crops could be separable in one area a different geographical area might not be suitable for monitoring by spectral sensors (e.g., tropical vs nontropical). A simple example of the computer's role in earth-resources features classification follows:

Example. *Study area:* farm fields.
Known number of crops: four.
Number of fields containing same crop: four.
Number of spectral readings taken per field: one.

From the above the total data that are available for analysis are
16 fields (four per crop type), each sampled once.

For crop type A four response curves (one per field) which may be *X-Y* plotted are shown in Figure 1. Note that two peaks occur in each of the four curves. More specifically, they occur at wavelengths λ_1 and λ_2 (Figure 2). When such mode commonality does occur, the next step is to generate a response plot in which the x axis reflects response measured at λ_1 and in which the y axis reflects λ_2 response readings.

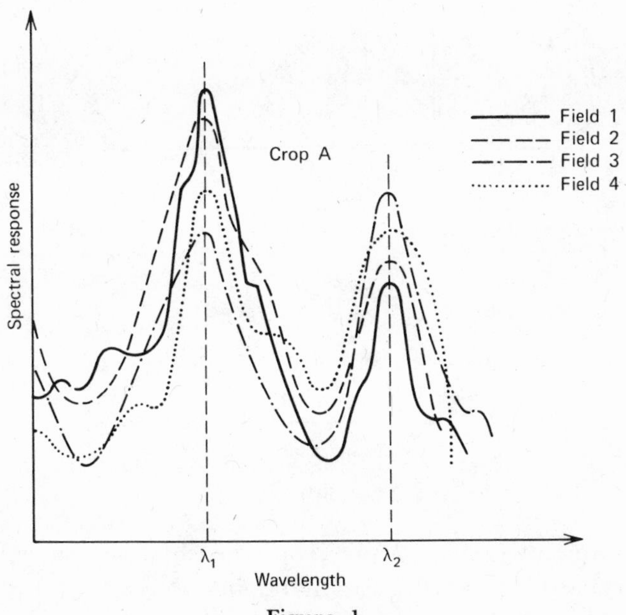

Figure 1

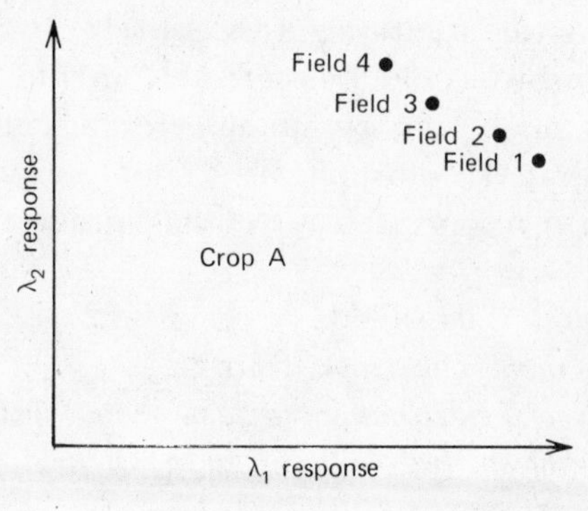

Figure 2

(Response consists primarily of reflected and emitted energy in a spectral region.)

Following the same procedure for crop types *B*, *C*, and *D* (in which four measurements per crop type are taken), we obtain Figure 3.

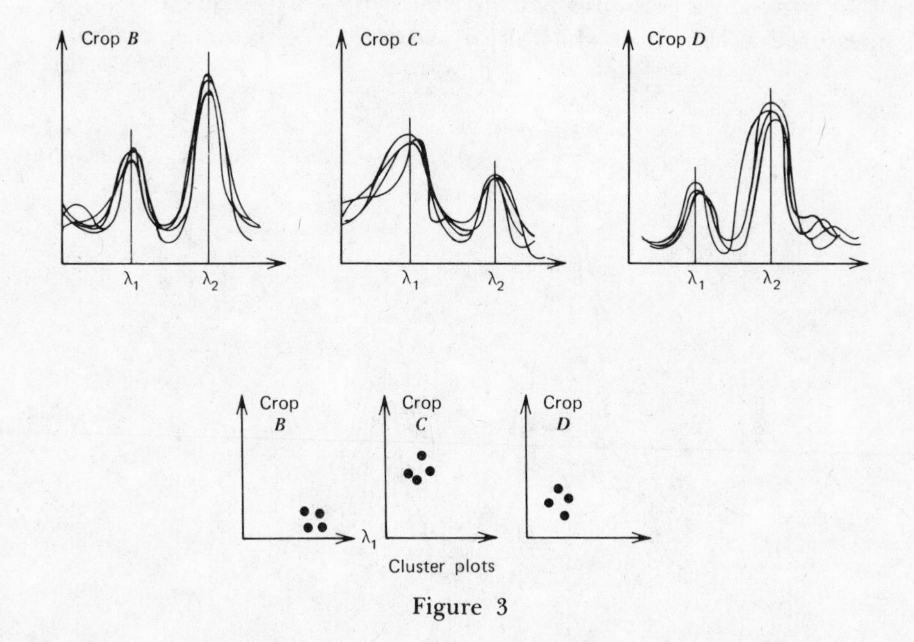

Figure 3

Having produced *cluster plots* per crop type, these may now be shown on a common *X-Y* graph. In addition, the digital computer (if properly programmed) may be used to determine cluster graph centroids and the locus of points equidistant from adjacent centroids. Examples are given in Figure 4.

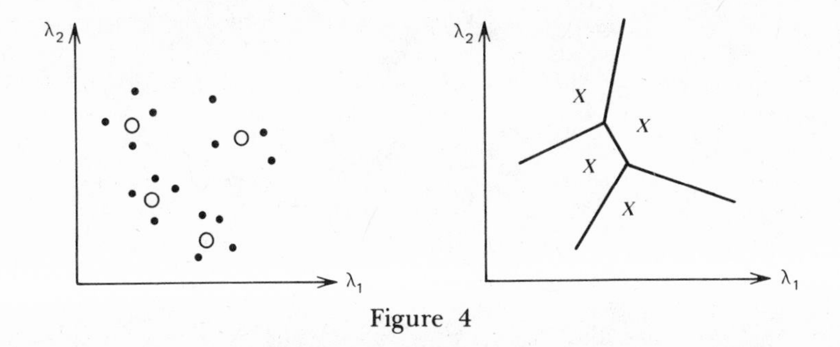

Figure 4

Suppose now that still another group of fields were samples and λ_1:λ_2 responses were shown as (1), (2) and (3) on the graph in Figure 5. Here point (1) can be classified as crop type *A*, point 2, crop *C*, but point 3 cannot be classified.

Pattern recognition forms the basis for multiband data analysis. Note, however, that areas of overlap in clusters can and do occur. Classification separation lines are also not always linear as shown in Figure 5.

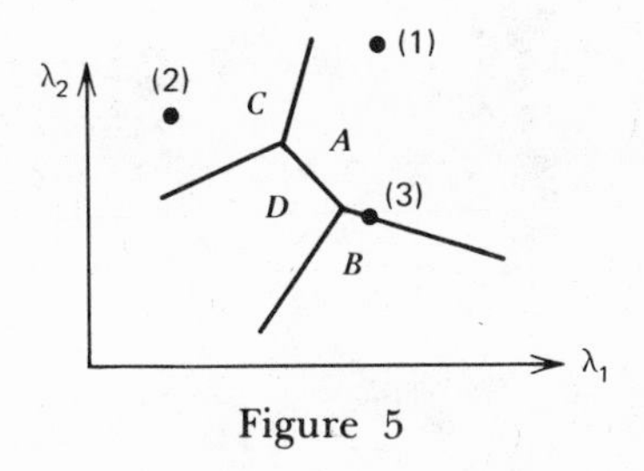

Figure 5

The accurate features classification program takes into account variance limits. In agricultural studies, for example, reference species (i.e., crops) must be assigned variance limits in terms of minimum and maximum plants per acre, anticipated plant area-to-exposed soil ratios, weeds per acre, and so on. A sufficient number of spectral readings must always be taken to ensure that a reference cluster contains points falling within variance limits. Limits that are chosen usually reflect conditions in the total test site to be surveyed by the aircraft or satellite multispectral remote sensor. For some features two readings are a sufficient measure of typical conditions throughout the test site. Other features may require three or more (Figure 6). When

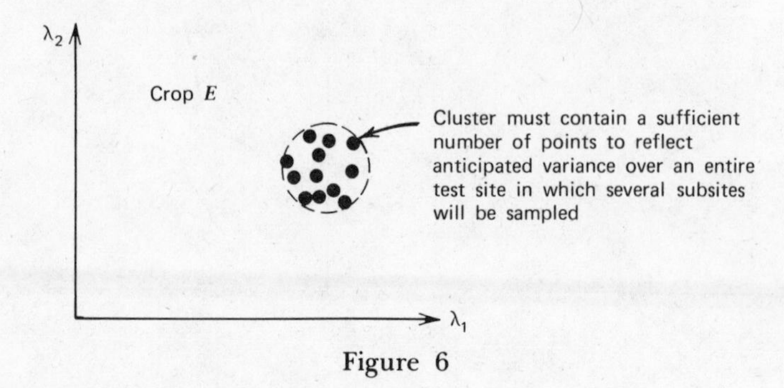

Figure 6

a specific cluster (e.g., Crop E) does not overlay clusters for surrounding vegetation, that feature is suitable for multispectral scanner monitoring and computer mapping via the λ_1:λ_2 technique (Figure 7).

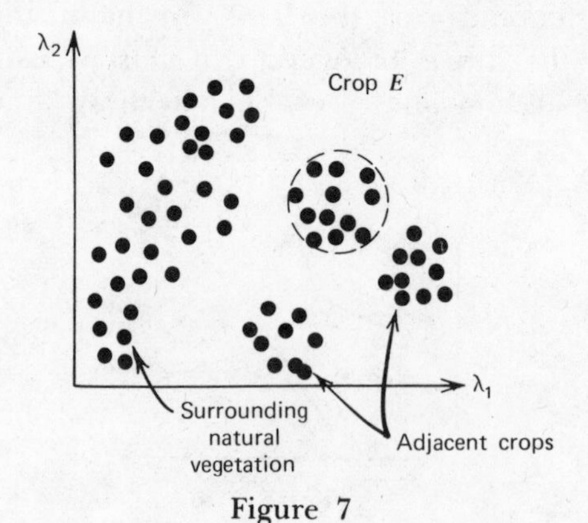

Figure 7

For certain applications λ_1:λ_2 response curves (i.e., two bands only) may not show features separation, whereas utilization of λ_1:λ_2:λ_3 graphs may offer better results, and comparisons at three wavelengths (λ_1, λ_2, λ_3) instead of two are often required. The λ_1 λ_2 λ_3 technique necessitates utilization of "3-D" graphics, as shown in Figure 8. To utilize the digital computer effectively for pattern recognition and automatic features classification, it is often required that reference sets of data be established. These reference sets, usually acquired from aircraft or ground survey, can be compared with new sets of multi-

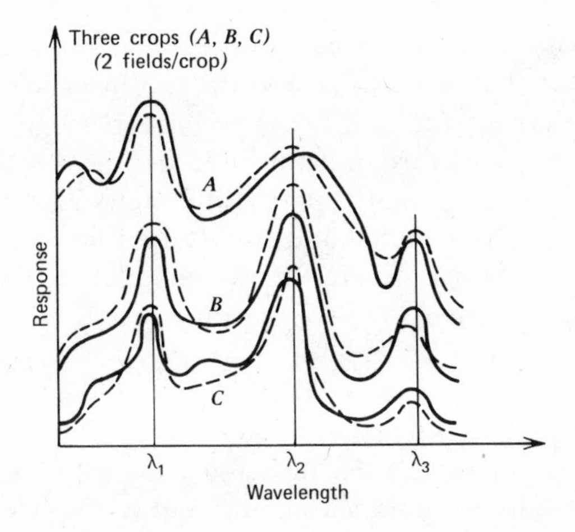

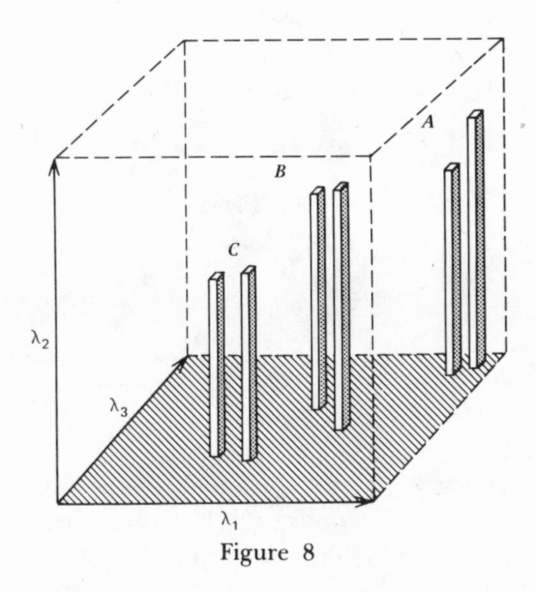

Figure 8

band data to determine the extent of correlation and whether pattern recognition is possible.

Reference sets of data, namely spectral signatures, are usually collected by using data already available (e.g., from photography) or by collecting ground truth signature data and multispectral scanner data for one or more features that are like others in a large scene.

Utilization of already available data is risky unless the trained data analyst can make adjustments based on expected differences in reference "historical" data versus the new data to be acquired. These differences can derive from one or more factors including variance in sun angle, ground moisture (in land studies), tidal conditions (in water quality studies), haze, and sensor viewing angle. This method is reliable only when environmental and man-caused factors, as well as sensor configuration, remain essentially the same for the time(s) at which historical data are recorded and the time(s) at which new data are acquired.

Establishment of reference signatures by locating an individual area (or several) typical of others throughout the test site and collecting multispectral sensor and *in situ* data for these reference scenes is an effective technique but time consuming and often more expensive.

Sometimes a lack of sharp definition between ground features causes a $\lambda_1:\lambda_2$ response curve to appear (see Figure 9), in which case a

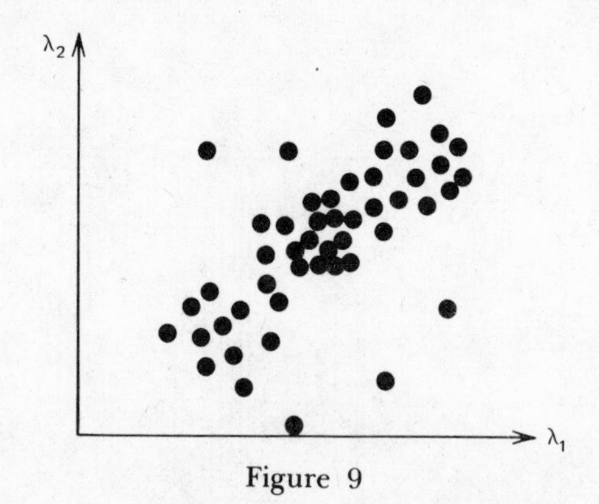

Figure 9

digital computer algorithm can be called on to perform clustering, that is, to define a fixed number of cluster centers and to associate each $\lambda_1:\lambda_2$ response point with a cluster or as being outside the boundaries of all established clusters. The number of clusters that software is instructed to construct may be based on the known number of feature types in the area under observation. If, for example, three predominant features were known to be in the area whose $\lambda_1:\lambda_2$ response curve was as shown in Figure 9, then an instruction to generate three clusters could lead to the scheme in Figure 10.

Determination of the validity of clusters generated from satellite data points near cluster edges, particularly at boundaries, requires valida-

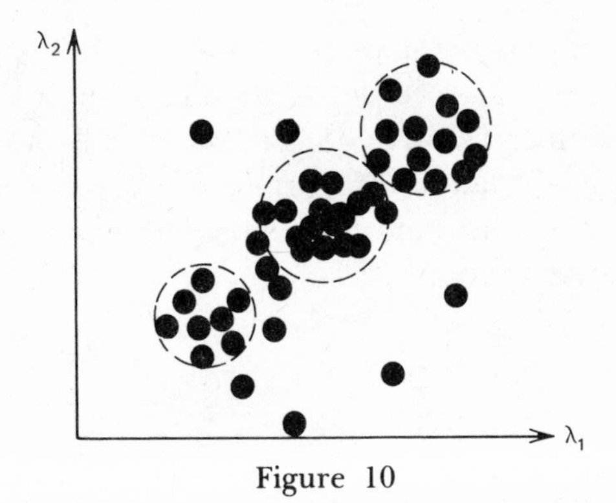

Figure 10

tion by field team(s) or by low-altitude aircraft data. When boundaries are verified, automatic classification/mapping software may perform the final step—earth feature mapping—shown later as a simple line print map.

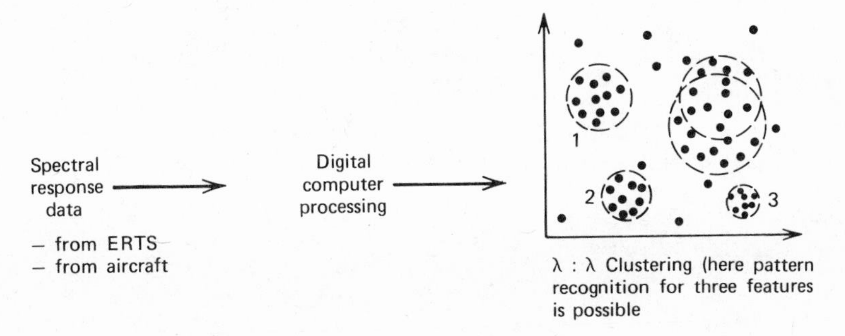

Figure 11. The mapping technique.

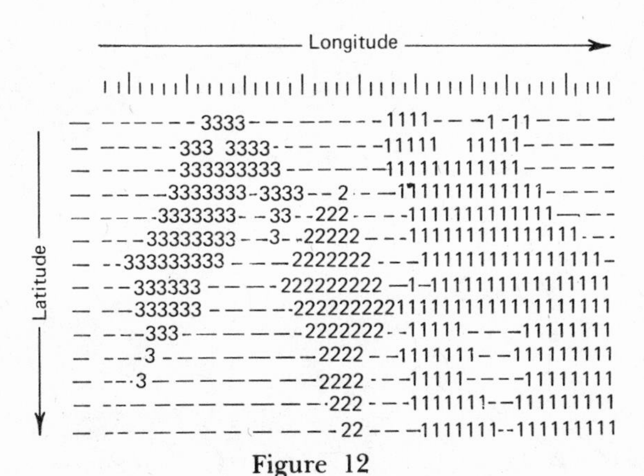

Figure 12

Figure 11 illustrates the mapping technique. The result is shown in Figure 12 as computer-processed scene data with automatic classification of three distinct features (1, 2, and 3); all others are (−). In addition to the basic latitude/longitude computer line-printer maps, gray scale shade, contour, and color mapping are possible. Map making and the basics of pattern recognition are discussed further in Chapter 15.

15

TECHNIQUES OF REMOTE SENSOR DATA PROCESSING AND DATA BASE DESIGN

Remote sensing entails the monitoring and classification of earth features by devices not in direct contact with the features under observation. When data of the aircraft or satellite multispectral scanner are acquired, the digital computer becomes an effective tool that effects geometric correction of data subject to sensor error and finally maps sensor results. Map is the key word; probably most useful to the land-use planner, environmentalist, hydrologist, geologist, and a host of others is the computer map. Essentially, the entire data-handling plan for a remote sensor program must ensure extraction of only meaningful data from the vast amount available from sensor systems and formatting them for map outputs. Final maps must be tailored to overlay standard maps or so that data from computer maps may be easily transferred to or compared with standard USGS, C&GS, or other maps.

With the pattern-recognition technique (see Chapter 14) the digital computer can "memorize" each unique spectral signature that pertains to soil and crop types, crop stress, water quality, and so on, and it can be rendered capable of generating map updates at rapid speed and high accuracy as additional or new data arrive. Also, one distinct advantage inherent in the computer is that data from two or more dates can be processed together to provide the analyst with a temporal analysis or time history of results. Frequently the overlay of data from multiple dates improves classification of features. A problem, however, in processing multiple scenes, is data registration. Care must be taken to ensure that, with respect to latitude and longitude, two or more scenes (or scene sections) overlay one another with acceptable accuracy.

Aircraft and satellite multispectral scanner data are similar. When they are analog, aircraft FM data may be analog-to-digitally converted; their outputs will be like satellite or aircraft digital data in that spectral measurements will appear in matrix format. Each computer-generated row of data generally contains spectral readings acquired from scan beginning to scan end. Row 2, for example, would contain data readings acquired during scan 2 of the sensor. The center columns in each matrix usually possess data acquired directly below the aircraft or satellite; the first and last columns of each matrix represent the extreme left and right measurements from each scan. Figure 1 is an actual example of satellite multispectral scanner data processed by a digital computer. Numbers to the left indicate scans made in the ERTS near-infrared spectral band. Each matrix character contains the detector reading for a 57 × 79m picture element or pixel (approximate dimensions).

Lowest energy values are coded with blanks and with the following symbols: "․," "+," "*,", and "%." The scene in Figure 1 is a portion of Puerto Rico's north coast—specifically the coastal area directly west of San Juan.

In displaying matrix data, numerous options are available to the data analyst. Generally, the first computer map produced shows a matrix with coded values, that is, one similar to Figure 1. A secondary product, and one experiencing widespread use today, is the gray scale shaded line-print map, in which high spectral values may be assigned a heavy-appearing typewriter character or a combination of two or more overprinted characters. Low spectral readings may be assigned light-appearing characters. Gray scale maps, when carefully executed, permit rapid interpretation of results with respect to standard coded line-print maps. When standard computer line printers are used, 10 to 24 different (and visually discernible) gray levels are appropriate. When electrostatic line printers are available, as many as 64 gray levels may be produced. Figures 2 and 3 are two examples of line-printer overprint maps.

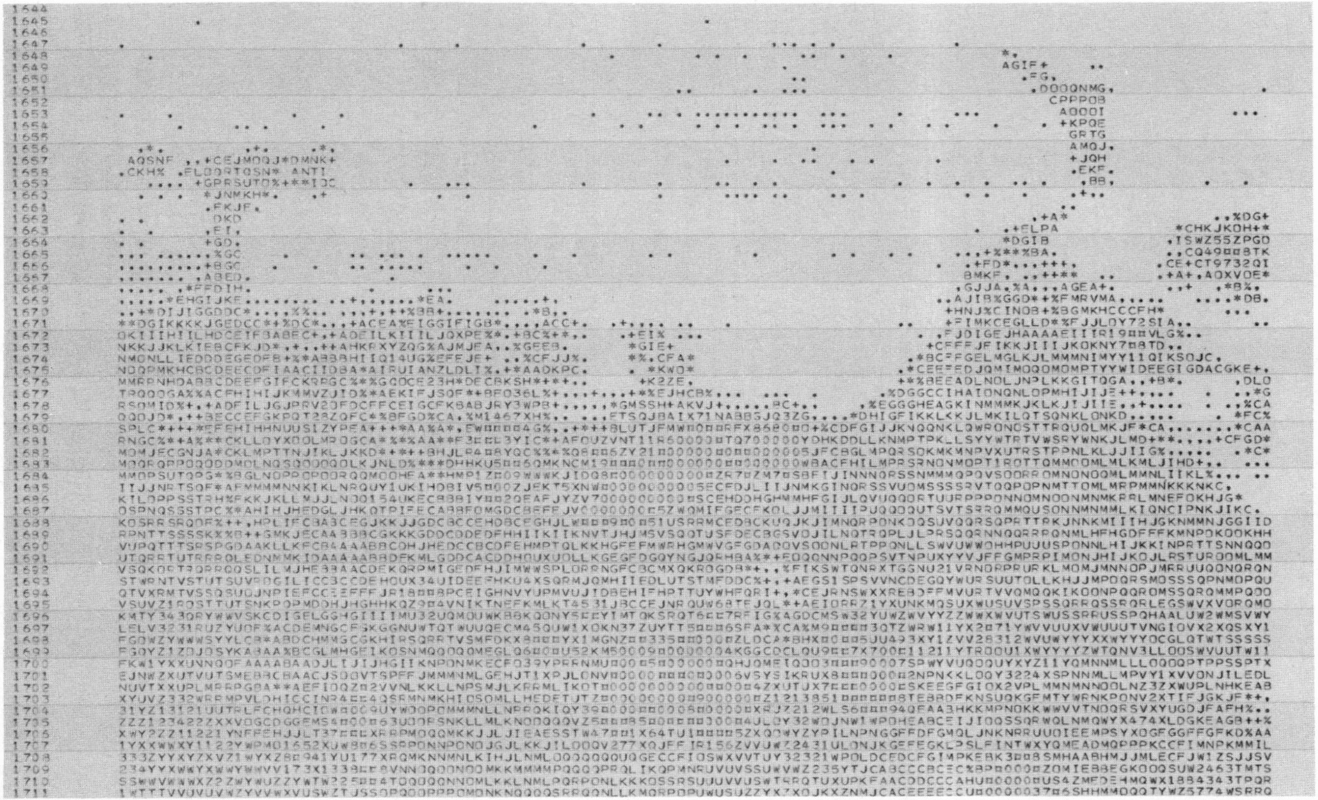

Figure 1. Computer line-print map, generated from ERTS data.

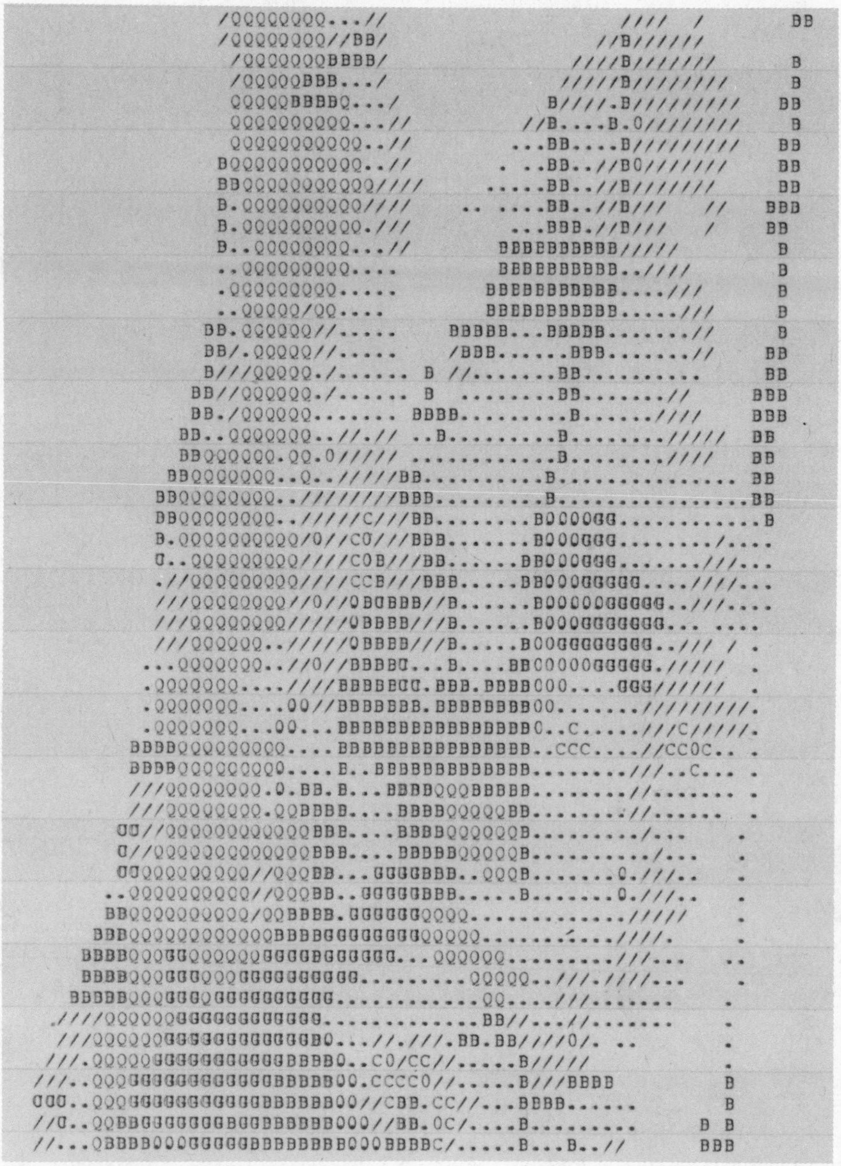

```
                /QQQQQQQQ...//                          ////  /        BB
                /QQQQQQQQ//BB/                        //B//////
                 /QQQQQQQBBBB/                      ////B///////      B
                 /QQQQQBBB.../                     /////B/////////    B
                 QQQQQBBBBQ.../                  B////.B//////////   BB
                 QQQQQQQQQQ...//               //B....B.0/////////    B
                 QQQQQQQQQQQ..//              ...BB....B//////////   BB
                BQQQQQQQQQQQ..//             . ..BB..//B0////////    BB
                BBQQQQQQQQQQQ////           .....BB..//B////////     BB
                B.QQQQQQQQQQ////        .. .....BB..//B///    //    BBB
                B.QQQQQQQQQQ.///            ...BBB.//B///    /      BB
                B..QQQQQQQQ...//           BBBBBBBBBBB/////          B
                 ..QQQQQQQQQ....            BBBBBBBBBBB..////        B
                 .QQQQQQQQQ.....           BBBBBBBBBBBB..../// 
                 ..QQQQQ/QQ....            BBBBBBBBBBBB...../// 
               BB.QQQQQQ//.....        BBBBB...BBBBB.......//     B
               BB/.QQQQQ//.....        /BBB......BBB.......//    BB
               B///QQQQQ./....... B //........BB..........       BB
              BB//QQQQQQ./....... B  .........BB......../        BBB
              BB./QQQQQQ....... BBBB..........B......////        BBB
             BB..QQQQQQQ..//.// ..B..........B........//////     BB
             BBQQQQQQ.QQ.0////// .............B........////     BB
            BBQQQQQQQQ..Q..//////BB...........B................  BB
           BBQQQQQQQQQ..////////BBB...........B..................BB
           BBQQQQQQQQQ..//////C///BB.........BQ00GG...............B
           B.QQQQQQQQQQ/0//C0///BBB........BQ0GGG........./....
           G..QQQQQQQQQ////C0B///BB......BBQ00GGG......../// ...
           .//QQQQQQQQQ////CCB///BBB.....BBQ00GGGGG..../////...
           ///QQQQQQQQ//0//QBGBBB//B......BQQQQGGGGG..///....
           ///QQQQQQQQ/////QBBBB///B......BQ0QGGGGGGG... ....
           ///QQQQQQ..//////QBBBB///B.....B00GGGGGGGG..///  /  .
          ...QQQQQQQ..//0//BBBBG...B....BB00000GGGGG.//////   .
          .QQQQQQQ..../////BBBBBGG.BBB.BBBB000....GGG//////   .
          .QQQQQQQ....QQ//BBBBBBB.BBBBBBBBB00......./////////.
          .QQQQQQQ...QQ...BBBBBBBBBBBBBBBBB0..C...../// C/////.
         BBBBQQQQQQQQQ....BBBBBBBBBBBBBBBBB..CCC....//CC0C..  .
         BBBBQQQQQQQQQ....B..BBBBBBBBBBBBBB......../// ..C...  .
         ///QQQQQQQQQ.Q.BB.B...BBBBQQQBBBBBB........//........  .
         ///QQQQQQQQQ.QQBBBB....BBBBQQQQQBB........//......   .
        GG//QQQQQQQQQQQQBBB....BBBBQQQQQQB........../...      .
        G//QQQQQQQQQQQQQBBB....BBBBBQQQQQB.........../...     .
        GGQQQQQQQQQQ//QQQBB...GGGGBBB..QQQB........Q.///..    .
       ..QQQQQQQQQQ//QQQBB..GGGGGBBB.....B........Q.///..     .
       BBQQQQQQQQQQQ/QQBBBB.GGGGGQQQQ.............../////     .
      BBBQQQQQQQQQQQQQBBBBGGGGGGGGQQQQQ........´..../// /.    .
     BBBBQQQGGQQQQQQGGGGBGGGGGG...QQQQQQ........../// ...    ..
     BBBBQQQGGGQQQGGGGGGGGGGG..........QQQQQ..///.////...    .
    BBBBBQQQGGGQGGGGGGGGGGG..............QQ..../// .......   .
   .////QQQQQQGGGGGGGGGGGGG...............BB//...//........  .
   ///QQQQQQGGGGGGGGGGGGGBQ...//.///.BB.BB////0/. ..         .
  ///.QQQQQGGGGGGGGGGGGGBBBBQ..C0/CC//.....B/////            .
 ///..QQQGGGGGGGGGGGGGBBBBBQQ.CCCC0//......B///BBBB          B
GGG..QQQGGGGGGGGGGGGBBBBBBQQ//CBB.CC//...BBBB.......        B
//G..QQBBGGGGGGGGBGGBBBBBBQQQ//BB.0C/....B...........     B B
//...QBBBBQQQGGGGGGBBBBBBBBBQQQBBBBC/......B...B..//      BBB
```

Figure 2. Gray scale computer map shows land-use patterns on eastern Long Island, New York.

```
       -+I==IA€€€€Z€M€€€€BBBBB€€MII++=+==++A==+---   1--   --    - -1+=+=IAA+1--1IXMZA
      -1+=XAAX€€€€€€€€€€€S€€€€€€€MA€++=+++++=111     --1       --1-   ---+  +--I=+=11X€€Z
     --=IAZXX€M€€€€€€€€€BBBBB€€€€MMAAAII===+=A+1-   ---111---   -1++1--1 -1+--+--+ZM€€€
    -1=AAAXAX€€€M€€€€BBBBB€€€€€€M€MXIIIII===X=111--+1-11111-   -1+++1---111-++IIA€€€€€
    -=IAXAAAXMMM€M€€€BB€€€€€€€€XEXIIX=IAI+I+111--1--11-    - -11-- -  +=+1-1=IXX€€€€B
    -+==IIIZZXZZ€€€€€€€B€€€€€€€€XIXAIXZI=XAI1++11111+     - 1+- 1     11+1    +-- 1=X€BB€€
   -1+===IIIXXXZX€€€€€€€€€€€€€€MXXAIIAIXZAII==1-- 1-   ----1-1111+I=++1      -+I=A€€BB€€
   --1+++==IAXIIX€M€€€€€€€€€€€X€XA=XXAXIAI==I++1          -11-11+++==II=1-     -+I€M€€€€BB€
    11+1=111+=IXXZ€€€€€€€€€€€M€€XXIIZZAII+I=11---111-------==++-1+==+-1--++X€€€€€€€€€€
  ---1+++++==IXXXA€€€€€€M€€€€€Z€IX+1ZAI=+++1-++-     -11-1IA+=11 1-1-++1-1+IM€BB€BB€€€B
     ----1-1=IIXZXM€€€€€M€€M€MZZZXA=AXXIII+=-1-   1-   1=AIA+=1----  --++1=A€€€€€€BBBA€
  -   111=IAXZMM€€€€€€€€€€€ZX€Z€ZAA=====1++=1   1111=II=I1+1--      ---11=AM€M€€€BBBMZ
         -1==ZM€€€M€€€€€€€€€ZAXAIXZA===11IAAI1-1++1-+=IAIII+=+-- -11-+A€€ZAZ€BSSBBB
          -AZM€€€€€€€€€€€€€€MMZAIAA€=A-1==+11+XXAAZX1=X===AI++111- -11=IA€€Z€€BBS€€€
           1X€€€€€€€€€€€€€€€€€€ZXAXIAIA+-1I€€=1+Z€€XXXZ€€ZIAAZZI=++11-=XXX€€€€€BBBB€€€
         11=Z€€€€€€€€€€€€€€€€€€€€MXZAAAI+IIIX€MZZAXM€M€€MM€MMM€MAA==++-=A€€€€BB€€BBB€MXI
         1=A€€€€€€€€€€€€€€Z€€M€€€Z€AAAZAAXXI=ZM€MXXMMM€M€€€€M€€MMZZAA=X€€M€€BBB€SBM--
          =X€€€€€€€€€BBB€€€€€€€Z M€X€AAZ€M€ZZMAIZ€M€MA€M€M€€€Z€€€€€ZZAIXM€€€BBSBBS€+
         1AM€€€€€€€€€€€€€€€€€€€€€XAI€€XZZ€ZA€€ZX=X€€€€X€€€€€M€€€MMZA=IAM€€BBSBBBBS€X-
        +I€€€€€€€€€€€€€€€€€€€€€€€ZXZZ€M€M€€Z€Z€€ZX€€€€€€€€M€M€€€€€€Z€XIIX€€€€€€BBSBB€€X1
        I€€€€€€€€€€€€€€€€€€€€€€€€€€Z€€X€€€M€€MZMMX€M€€€€€€€€€€€€€€€X€€€€€€€€BBBBSBB€M=1-
       1€€€€€€€€€€€€€€€€€€€€€€€€€Z€MZZZ€€€€€€€€M€XMMM€€€€€€€€€€€€€€€MM€€€€€€€€€BBBSBB€€A-
      -=€€€€€€€€€€€€€€€€€€€€€MM€€€€M€MM€€€€€€€€€€€ZM€€€BBB€€€€€€BBBB€M€€€€€€€€€€BBSSB€M+
     +XM€€€€€€€€€€€€€€€€€€€€€€€€€€€€€€€€€€€BBB€€€€€€€€€€€€€€€€€€€€€€M€€BBB€€€€BBSBB€€+
     1Z€0€€€€BB€€€€€€€€€€€€BBB€€€€€€€€€€€€€€€€€€€€€€€M€€€€€€€€€€€€€€MZZZ€€BBBBBBBBB€€€Z+1
    =X€€€BBBBB€€€€€€€€€€€€€BBSB€€€€€€€€€€€€€BBB€€€€€BBBB€€€€€€€€€MM€€€€BBSSSBBBB€1--
    1I€€€€€BBBBB€€€€€€€€€€€€BB€€€€€€€€€€€€€€BBB€€€€€€€BBB€€€€€€€€€ZIIX€€€€€€BBBSBBBB€M+
    Z€€€€€€€BBBBB€€€€€€€€€€€BBB€€€€€€€€€€€€€€€€€€€€€€€€€€BSB€€€€€€€€€€€€€ZA€€€€€€€€€€€€€€€=
   1€€€€€€BBBBBBBB€€€€€€€€€€€€€€€€€€€€€€€€€€€€€€€€€€€€€€€€€€€€€€BSBBB€€€€€€€BBSS€€AAM€€€€€€€€€€€€€+
   I€BBBBB€BBBBB€€€€€€€€€€€€€€€€€€€€€€€€€€€€€€€€BBBBBBBBBBBSBBBBSSBBB€M€=1=IAAM€€€€€€€€X1
   Z€€€BBBB€€BBBBB€€€€€€€€€€€€€€€€€€€€€€€€€€€€€€€€SSBBBBBBBBBBSBB€€€€€€I1=IX€€€€€M€XI-
   A€€€€€€€BBBBB€€€€€€€€€€€€€€€€€€€€€€€€€€€€€€€€BBBBBSBBBBBBBBBSBBB€€€€BBB€€M€€€€€€€XA=-
   =€€€€€€€€€BBB€€€€€€€M€€€€€€€€€€€€€€€€€€€€€€€€€BBBBSSSBBBBBBBBBSBBB€€€€BBBBBBBB€€€€€€€X+-
    X€€BBB€€€€€€€€€MM€€€€€€M€€€€M€€€€€€€€M€€€BBBSBBBBBBBBBBBBBBBBB€€€€€€BBBBBBBB€€€€€+
    Z€€€€€€€€€€€€€€€€M€€€€€€ZZ€€€€Z€€€€€€€€€€€€€€BSSSSBBB€€BBSSBBBS€€MZM€€€BB€€€€€€€MZI
   1€BBBB€€BBS€€€€BB€€€M€€€ZMXAAXIA€€€€€€€€€€€€€€BBBBBSBBBBBBB€€BBSSBB€€MZZZ€€€€€€€MXAZX+
   €BSSBB€€€BS€€€€€€€€M€€€AZI=X=++€€€€€€€€€€€€BBBBBBBBBBBBBBBBBBBBBB€€€€€€M€€€€€€€€Z1=A-
   €BBB€€€€€BS€M€€€€€€€€€€MIIZ1ZI-1=ZBBB€€€€€€€€SBBBBBBBBBBBBBB€€€€€€€€€€€€€€€€€€€€€M1-1-
   €BBBB€€€€SS€€€€€€€€€€€€€€A+€IIX-  ++M€€€€€BBB€€€€€BB€€€BB€€€€€€€€€€€€€€€€€€€€€€€MM+
   €BSSSB€€SSS€€€€BBBB€€€€€€Z=XX+A+11=IM€€€€BBBB€€€€€€€€€€€€€€€€€€€€€€€€€€€€€€€€M€ZA
   1€BBSSSSSBB€€€BBBBBB€€€€€€€€X=XI+I+1=IAMM€€€€BB€€€€€€€€€€€€€€€€€€€€€€€€€€€€€€€€€AA1
    =€€€€BBBBB€SBBB€€€€€€€€€€€€I=€+I1++XZZ€€€€€€€€€€€€€€€€€€€€€€€€€€€€€€€€€€€€€€€ZI1
      I€€€€BSBBBBB€€€€€€€€€€€MZIA€I=1+=X€M€€€€€€€€€€€€€€€€€€€€€€€€€€€€€€€€€€€€€€Z=
        1AM€€BBBB€€€€€€€€€€€€€€I1+IAIIXX€M€€€€€€€€€€€€M€€€€€€€€ZZ€M€€€€€€€€€€€€I
           -=Z€€€€€€€€€€€MXZA+1+A€€M€M€€€€€€€€€€€€€€M€M€ZM€Z€€€€€M€€M€M+
             -IM€€€€€€€€€ZAI+=+I€M€€€€€€BBB€€€€€€€€€€€ZZZ€M€Z€ZA=Z€€ZZ1
               +Z€€€€€€€€€€€€€€€€XA€M€€€€€€BBB€€€€€€€ZXZAXXX€€€€€XI+1XMI=1
                -=AM€€€€€€€€€€€€XAZM€€€€€€€€€€€€€€€€€€€AAAAXX€€€€€Z=+=IA+1
                    1I€€€MAAAAIAZ€€€€€€€€€€€€€€€€€€MXAAAAZ€€€€€€€ZXXA=+1-
                          -=I==IX€€€€€€€€€€€€€€€MMZXAX€€BBBB€€€€€€€€ZAAI1
                           -====AX€€€€€€€€€€MMMMM€Z€M€€€€€€€€€€€€€€€ZI
                            -++==IAZ€€€€€€€MXM€ZZ€€€€€€€€€€€€€€€€€€€ZXAX1
                               -1=++=A€M€M€€€XX€MM€Z€€€M€XXZXI==+
                                 -1 11=IXZXXXXA€€M€XAAI=11--+=+=1-
                                  -     -11=IAIIXX=-
                                             1+++I=-
                                                11
```

Figure 3. Gray scale line-print map (16 shades).

Note that Figure 3 was generated by symbols different from those used in the Long Island map. Line-printer gray scale maps can be produced at low cost and are easy to reproduce on almost any office copier.

It must also be noted that an extremely valuable (but expensive) tool for displaying scanner data is the color cathode ray tube. Instead of converting spectral values to gray shades, we can create color presentations instead. Like the gray scale map, color presentations are, in effect, similar to low-resolution photographs and are useful to analysts in all study areas. Also commercially available are devices that permit conversion of digital data to colored ink on paper displays.

One significant problem, however, with the basic computer line-print map is that there are 10 character spaces left to right and 6 (or 8) vertical characters spaces in a square inch of computer printout (Figure 4). Thus original scanner data are not presented in square

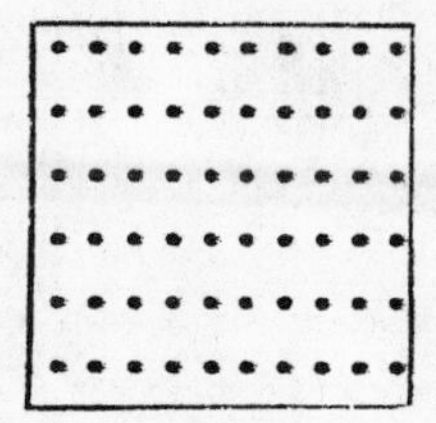

Figure 4

arrays but are distorted into rectangle. If, by photo processing it were possible to stretch printouts horizontally or to compress them vertically, square matrices would result and, with photo-enlarging equipment, they could be made to overlay standard maps. This, however, is not possible. For satellite data, specifically from the multispectral scanner, this problem is even further aggravated because each matrix cell or pixel (picture element) represents a ground area that is not square. Each Resource Technology Satellite (ERTS) computer printouts, for example, contain pixels that represent energy recorded in areas of about 57 × 79 m (Figure 5).

Obviously these data cannot overlay a USGS or other standard map. By programming, however, data may be made map compatible. One technique involves rotating a data matrix, deskewing it, and rescaling it to meet map requirements (e.g., scaled for 1:24,000 or 1:20,000 type USGS. or C&GS maps). When computer data-to-map overlay is essential, it is advisable to use geometric correction and scaling computer programs.

An alternate manual approach entails the construction of a rectangular grid over a map and the transferrence of pixel values into the new grid. This manual (drafting) technique proves to be effective when only small quantities of data are to be analyzed. Figure 6 is an ERTS grid (i.e., 57 × 79 m cells) overlayed onto a standard map of St. Thomas.*

*Source. NASA ERTS-1 Experiment #589, October 1973.

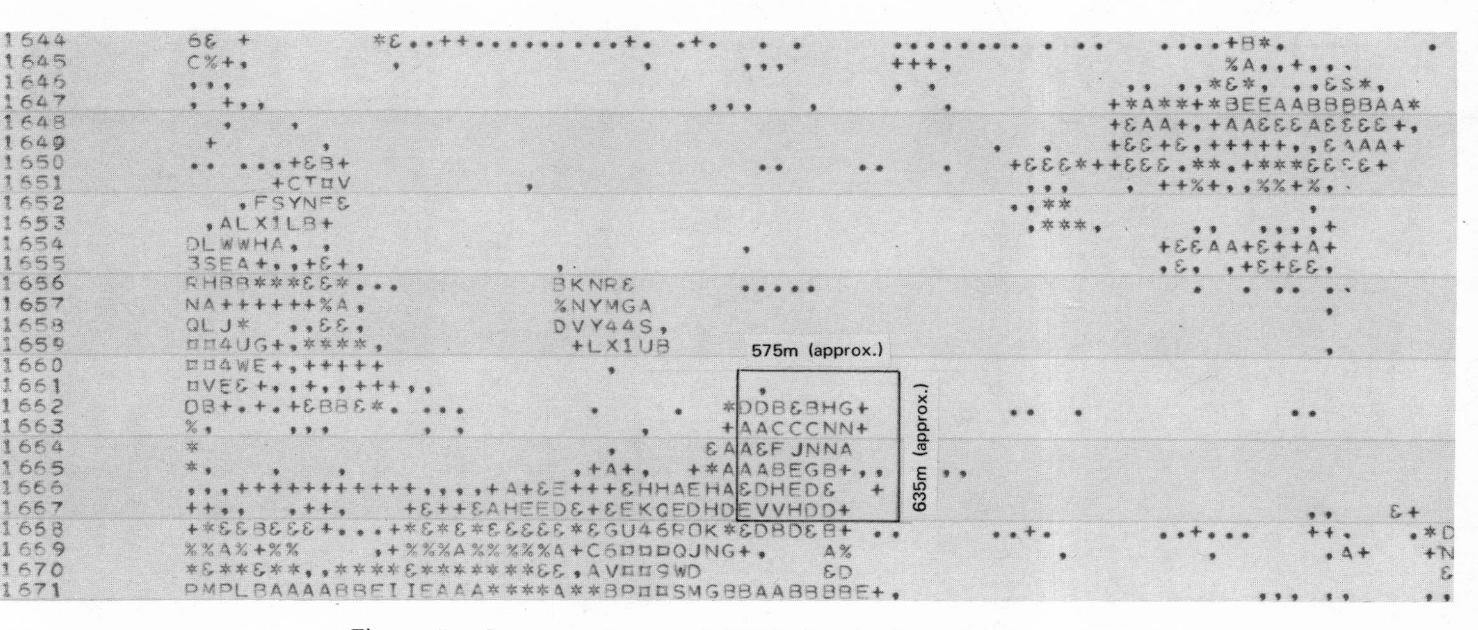

Figure 5. Computer-generated ERTS data in line-print format.

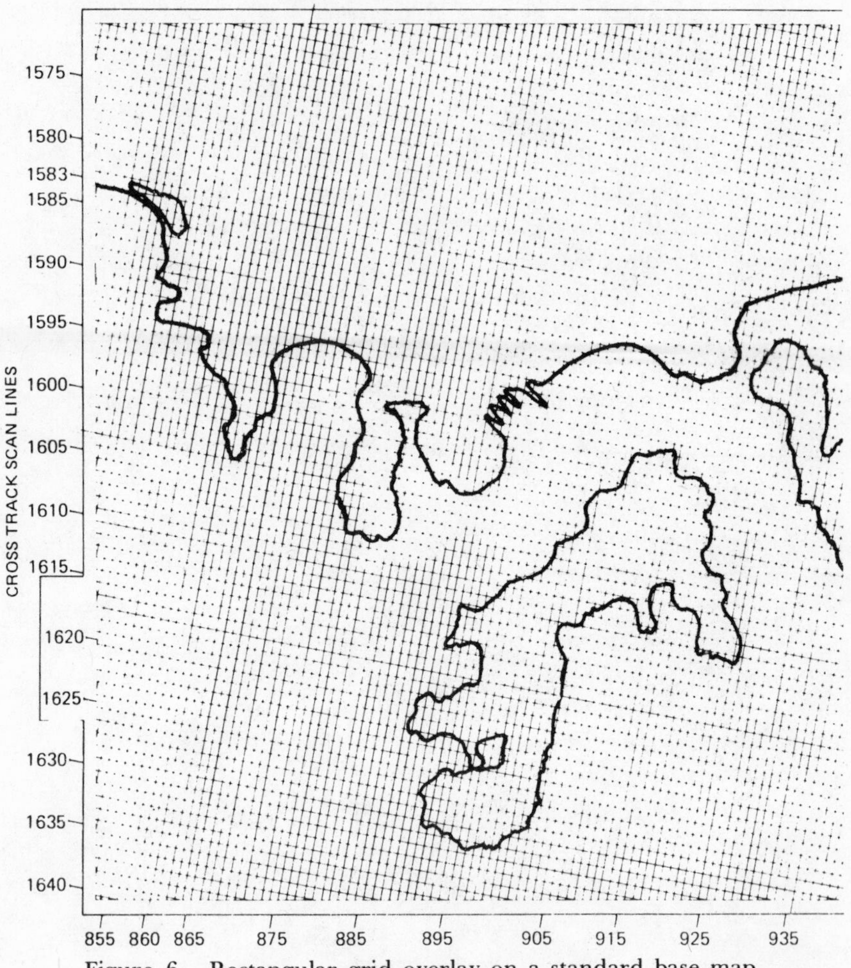

Figure 6. Rectangular grid overlay on a standard base map.

Also noteworthy when limited quantities of data printouts are needed, are the color presentations generated by use of standard artist pencils. Although this "color-by-the-numbers" exercise requires several hours, the results are certainly suitable for presentation, for they have the same or better resolution as map displays generated by expensive computer-driven color cathode ray tubes. An additional point for consideration is that although a 10 to 24 color shade map is more attractive than a 10 to 24 level gray scale map the cost of duplicating color products for reports or presentations far exceeds that of black and white.

Once the problem of scaling has been overcome, gray scale line-print maps or color products may be compared with known data (ground truth), and when high sensor-to-ground truth data correlation is achieved the computer can go on to recognize and map similar features in a given date's data or for data acquired at a later date.

Still another technique, which permits later mapping by computer, first requires clustering (discussed in Chapter 14). Here, specialized computer programs let the computer digest data from several spectral bands (e.g., 3 or 4) and the machine is instructed to obtain clusters or groups of data that are spectrally separable from one another for a matrix area of interest. A request, for example, for the generation of 10 to 15 clusters from ERTS data is typical. By reviewing the mapped results the analyst can determine patterns that correlate with known facts; when confirmed correlation is established, appropriate patterns are designated as *training sets* and used in the computer to map new or additional data. Up-to-date aircraft photography is generally extremely useful in establishing the degree of correlation between clusters and ground features and should be employed when available. In this *nonsupervised classifier system* the analyst usually has the option of trying fewer or more spectral bands and/or of asking for a different number of clusters. Once training sets have been established, follow-on pattern recognition and computer mapping fall into an operation commonly referred to as *supervised classification.*

After he has completed clustering and associated computer mapping it sometimes becomes apparent to the analyst that other features not of direct interest to him are also mappable from aircraft or satellite scanners. If he is responsible for water-quality mapping, for example, he might find that certain coastal vegetation types (e.g., mangroves and swamp grass) are effectively monitored from above.

The foregoing paragraphs are intended to serve basically as an overview of modern environmental or earth-resources data-handling techniques as they stand today. It must therefore be pointed out that technology continues to advance in the pattern recognition/computer mapping field and that, depending on the nature of the geographical region and features being studied, it can become advisable to exercise specialized computer programs to enhance or to make more accurate final computer results. Use, for example, of functions that include the mean, standard deviation, correlation, and covariance can permit the establishment of a better spectral fingerprint for classes under observation. Divergence analysis programs may at times be utilized to help pinpoint the spectral bands, from a group of four or more, that are most appropriate in feature detection and mapping. Histograms, Gaussian maximum likelihood classifiers, and correlation coefficient matrices are other tools that can indeed prove to be helpful.

As the field of remote sensor data handling advances through the 1980s, each application area, for example, water-quality and crop monitoring, will have associated with it recommended spectral monitoring bands and computer classification and display programs. Basic tools for all applications, which include the line-print map, the gray scale map, color presentations, and clustering algorithms, are here today. Advanced tools have in many areas been tested with success and some operational procedures, particularly for agriculture, have been established.

For the comprehensive survey or monitoring program, however, strong emphasis is placed on the computer and graphics-display techniques. Also, in the early stages of remote sensing programs, large quantities of ground truth data are frequently essential. Whether acquired by boat, plane, or field crew, these programs are often best handled by storage in a specially tailored computer data bank (or data base) from which maps, listings, and the data required to establish the boundaries of application and specific degrees of correlation between satellite and ground truth data may be extracted.

DATA-BASE DESIGN CONSIDERATIONS

One major factor in programs involving environmental or test data handling is the proper formatting of all acquired data so that, from a data bank, a data analyst can easily extract final computer-processed results in map or graphics summary form.

As mentioned earlier, typewriter listings of all acquired data, whether from the transducer or remote sensor, are rarely as valuable as graphics presentations. Maps and graphics summaries are easily comprehended by the data user. Data anomalies or out-of-expected-range readings are usually best given graphically. Thus in the design of an overall data processing plan it is vital to structure the data base so that its contents can be processed for the graphics or listing format. In effecting the graphics presentations, test data are usually plotted in

X-Y format, that is, measurement value versus time; earth-resources data are generally placed in map format in which sensor values are shown on a graph whose x axis is related to longitude and y axis relates to latitude. For those time zones or geographical areas that contain anomalies or data that are appropriate for more detailed analysis, the computer can then be called on to "zero in" on these areas and list all available data for this time or geographic zone. By the graphics products initial use, all data can be efficiently reviewed. (It is after this point that computer line-print listings are generally required.) Without graphics, analysts are frequently inundated with pages and pages of typed results and zeroing in on regions of interest is extremely time consuming.

Thus a major consideration in the design of an overall data-handling plan is the formatting of input data (from transducer or sensor) so that data from data banks can first be displayed graphically and then retrospectively listed for each time or geographic region of interest.

In earth-resources data handling data should be formatted for later presentation on a base map that overlays or can be visually correlated to the USGS, C&GS, or other standard map. One method of generating a computer base map involves entering data related to map features into a computer, programming the machine to generate a base map that resembles the original, and displaying *in situ* or remote sensor data.

As an actual example a Coast and Geodedic Survey Map of St. Thomas Harbor was digitized and later presented, by computer, in a format by which each typewriter data cell represented an area with dimensions of 1 sec latitude by 1 sec longitude (Figure 7). Onto this map, that is, in the water areas, the results of *in situ* data or satellite data could then be presented. Shown in Figure 8, for example, is a graphics summary of *in situ* temperature readings taken over a six-week period. Clearly, the computer graphics technique offers the analyst a means of readily relating computer-processed results to the geographic area under study and of reviewing a map rather than numerous lines of typed statistics.

Creation of the base map may be achieved in several ways: when only one or a low number of maps is involved, manual methods work well. A grid is superimposed on the original map(s) and key features are recorded, by code onto 80-column keypunch cards. Cards are

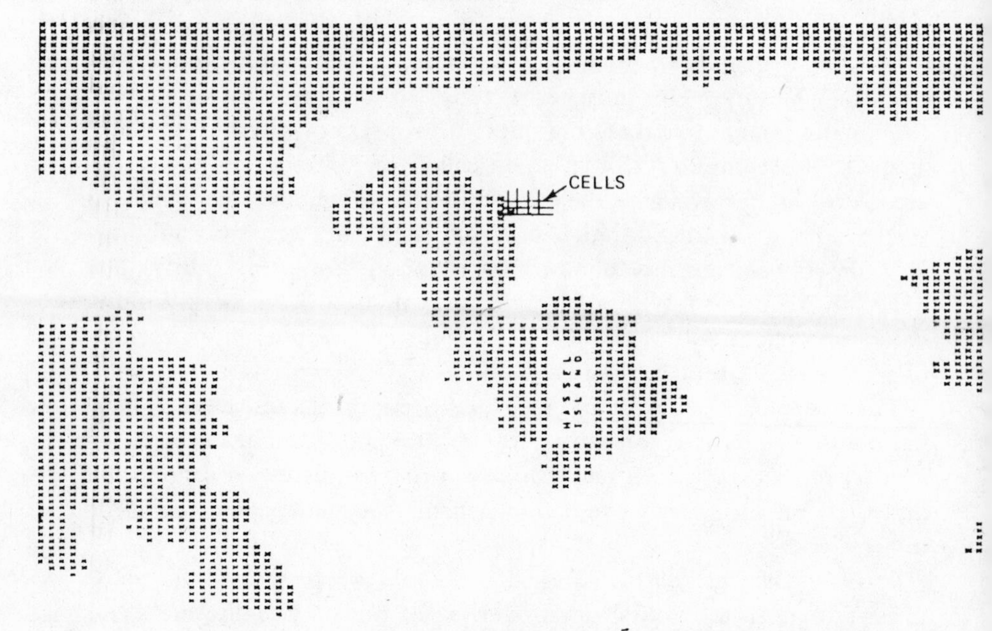

Figure 7. Computer base map, St. Thomas harbor.

formatted to contain the measurement code and other information related to cell position in the overall latitude/longitude grid. In water-quality surveys land areas may be coded with the same symbol or with a combination of symbols perhaps to reflect the location of key control points such as roads or unique natural features. Water areas are left blank. For land surveys it is recommended that only those key features be coded that are not directly at points at which ground or remote sensor data are to be acquired; for example, when crop surveys are to be made, roads can be coded with *R*, water bodies with *W*, and other features with other codes. Manual coding is most time consuming, and it is important that the minimum number of points be coded to render the final line-print map directly comparable to the original.

A second technique in base-map generation requires the use of a semiautomatic digitizer. Here map-to-computer card or map-to-computer tape conversion is achieved by placing the digitizer's cross hairs over the point to be digitized and depressing a button. In the sonic digitizer a follower pen is placed on the point of interest and pressed down each time a point is to be digitized. For each "press" a

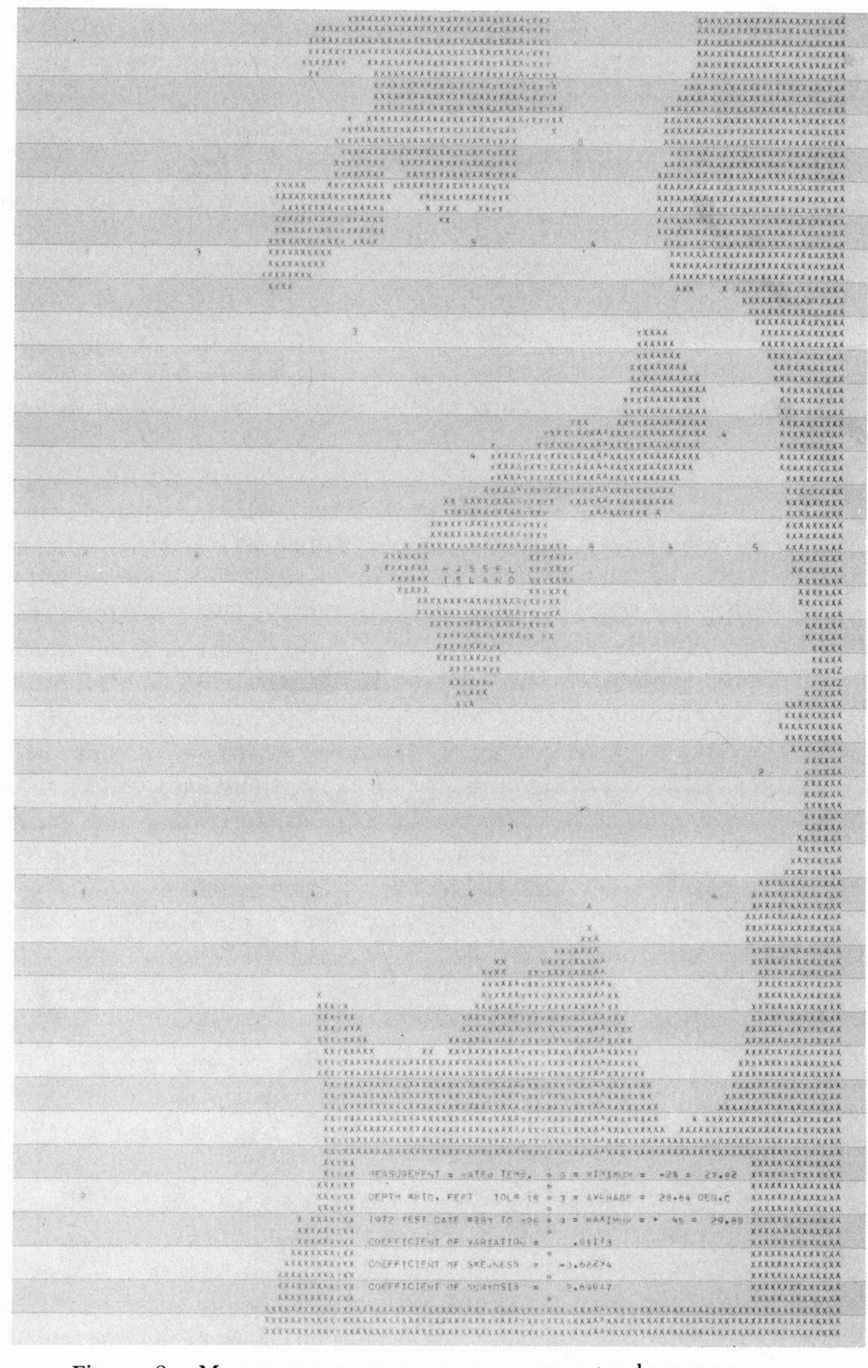

Figure 8. Measurement summary on a computer base map.

card is created or an entry is made on magnetic tape. Before each unique feature is digitized, a code is selected, which during digitizing is printed with the *x-y* location in which it occurred. Digitizing rates of 30 to 40 points/min are typical of some applications.

No matter what the technique utilized, each converted set of base-map data must have coordinates associated with it that relate to coordinates of the original map and a coordinate format like that for acquired *in situ* or remote sensor data. When large quantities of base maps are to be generated, a digitizer is usually a best approach. When manual methods are employed, care must be taken because of the lack of digitizing hardware or only when a small quantity of map data is used, for manual digitizing has a higher probability of error occurrence.

As a final note, while still on the topic of digitizers, mention is made of another commercially available digitizer, that is, one that projects microfilm or aerial film images onto a reading screen. It, too, contains the cross-arms follower and cross-hairs viewer.

Data-base design for the environmental data-handling program essentially requires the establishment of a graphics-oriented information management and retrieval system. The system chosen must be flexible enough to permit rapid updating of stored records, retrieval of data in graphics or listing format, and access to records by sections or by conditions criteria. Conditions, for example, could include calling for data only within user-specified time periods, for listings (or graphs) only when measurement values exceed or are lower than a specific value assigned by the analyst, or for data when some combination of conditions exists.

A large number of commercially available data-base software packages are ready for installation on the analyst's computer. If a computer is not obtainable, the analyst may rent (or buy) a computer terminal which can be linked by telephone lines to another machine.

Essentially, the actual computer data base is a matrix in which, for each record (specified in columns $1 \rightarrow n$), there is a series of measurement characteristics, each listed in specified fields or column positions beginning at $n + 1$. Measurement characteristics in the well-designed system can be recorded as alphanumeric notations, integer numbers, or real numbers.

Certainly, before developing or procuring a data-base system all details pertaining to the nature of data to be acquired must be known. With this knowledge it is then recommended that *in situ* data recording logs be structured with their contents in machine-readable form and with record and field positions identical to those of the data base. Having done these things, boat logs, logs generated by field crews, and laboratory analysis logs need only be keypunched; the cards can then be entered (as files) in the data base. Also, the analyst can, at a terminal typewriter, key in pertinent log sheet results.

When the data bank is established, the data-base system will allow the user to make new entries or deletions, call for all or parts of the bank based on his conditions, and generate final computer reports in the graphics or listing format. In large computer systems, generally, no limit is imposed on the size of the data-base matrix, and in the modern machine a variety of graphics display options for data extracted from the data bank includes line-print graphs, computer-output microfilm, and the computer-output *X-Y* or *X-Y-Y′* plotter.

In the environmental data base all measurements are usually referenced to map coordinates. Thus, once data are extracted on the basis of conditions imposed by the analyst, it is a relatively simple programming task to present data on the computer base map. In Figure 9 data were removed from a data bank for parameter = water temperature, for conditions depth = 5 ± 1 ft, and for Julian days 283 to 322. Once removed, another computer program averaged all data available within the test zone and presented scaled data values per sample station. When graphics presentations point out an area worthy of more detailed study, the analyst may then extract comprehensive listings for parameters or regions of interest. The listing in Figure 10, for example, is an actual data-base output for data acquired during one day and at one location. Finally, it must be emphasized that for success in an environmental data-handling program it is advisable to bring together at the start of a project all data processing personnel and data acquisition and analysis team members. Each should participate in the development of an overall data-handling plan.

The well-executed plan will permit the extraction of meaningful data only and can prevent inundating the team of analysts with pages and pages of results, many of which are of little use in the application of data to problems or studies that must be addressed.

Use of one or more of the numerous graphics options available is recommended; many graphics displays are dramatic in appearance and present both qualitative and quantitative results on a single page. Selected examples are given in Figures 11, 12, and 13.

In the early stages of programs, when ground truth data are ac-

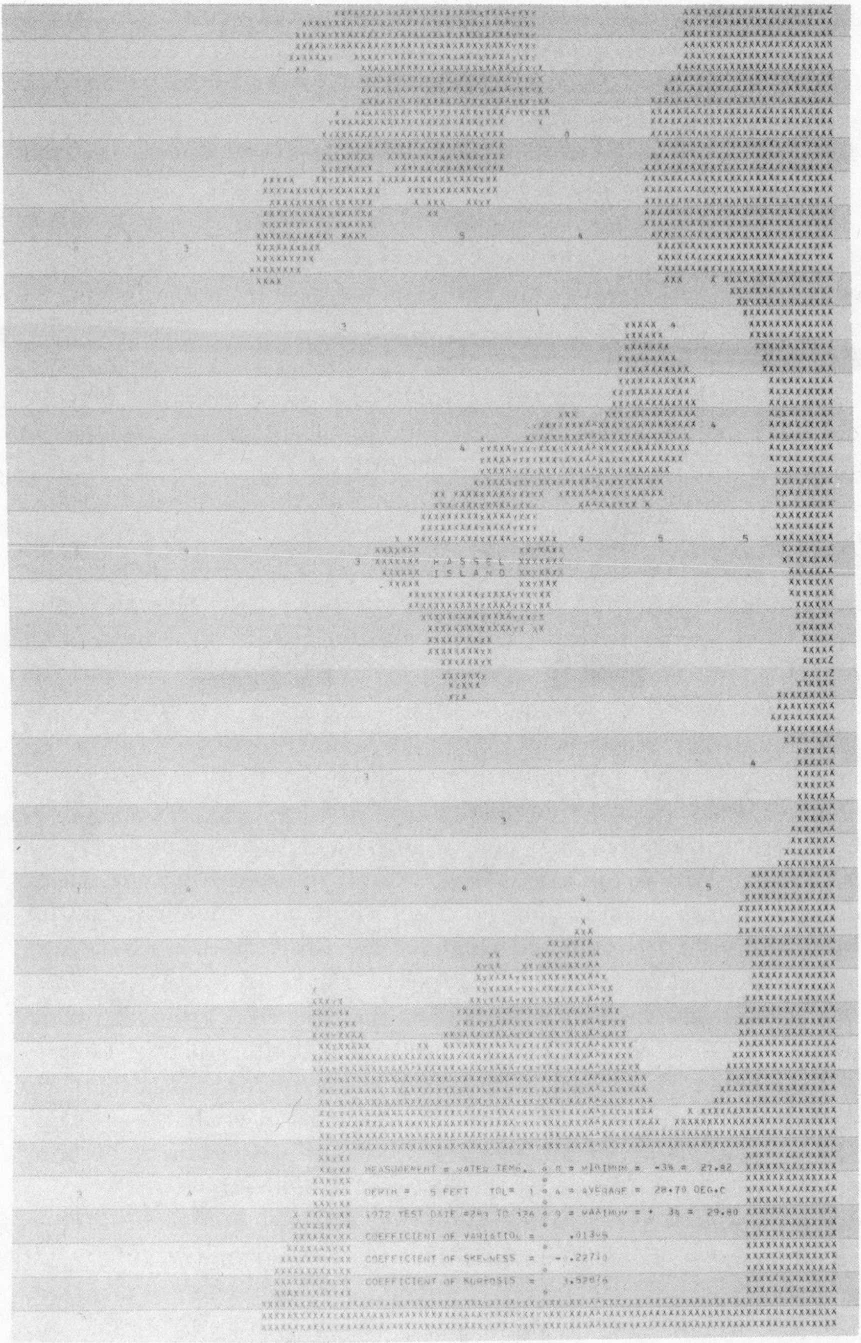

HASSEL ISLAND

MEASUREMENT = WATER TEMP. * 0 = MINIMUM = -3% = 27.82
DEPTH = 5 FEET TOL= 1 * 4 = AVERAGE = 28.70 DEG.C
1972 TEST DATE =283 TO 326 * 9 = MAXIMUM = + 3% = 29.80
COEFFICIENT OF VARIATION = .01345
COEFFICIENT OF SKEWNESS = -.22710
COEFFICIENT OF KURTOSIS = 3.52876

Figure 9

VIRGIN ISLANDS TEST SITE PROGRAM

ST. THOMAS HARBORS WATER QUALITY DETERMINANTS STUDY

STATION = 9
TYPE = 3

BLOCK	DATE TIME	WATER DEPTH FEET	SAMPLE DEPTH FEET	DIS.OXYGEN PPM	--PH-- NONE	CONDUCTIVITY MHO/CM2	SALINITY PPT	WATER TEMP. DEG.C
1218	325 1000	32.00000	27.00000	5.60000	8.20000	56.90000	36.00000	27.00000
1224	325 1100	29.00000	5.00000	5.70000	8.20000	57.00000	36.00000	27.90000
1225	325 1100	29.00000	15.00000	5.80000	8.20000	57.00000	36.00000	27.90000
1226	325 1100	29.00000	24.00000	5.70000	8.20000	57.00000	36.00000	27.70000
1232	325 1200	29.00000	5.00000	5.80000	8.20000	57.00000	36.00000	27.90000
1233	325 1200	29.00000	15.00000	5.80000	8.20000	57.00000	36.00000	27.90000
1234	325 1200	29.00000	24.00000	5.80000	8.20000	57.00000	36.00000	27.90000
1240	325 1300	29.00000	5.00000	5.90000	8.15000	57.00000	36.00000	27.90000
1241	325 1300	29.00000	15.00000	5.85000	8.18000	57.00000	36.00000	27.80000
1242	325 1300	29.00000	24.00000	5.85000	8.17000	57.00000	35.90000	27.80000
1248	325 1400	29.00000	5.00000	5.90000	8.17000	57.10000	36.00000	27.90000
1249	325 1400	29.00000	15.00000	6.00000	8.18000	57.10000	36.00000	27.90000
1250	325 1400	29.00000	24.00000	5.95000	8.19000	57.10000	36.00000	27.00000
1256	325 1500	29.00000	5.00000	6.00000	8.18000	57.10000	36.00000	28.00000
1257	325 1500	29.00000	15.00000	5.99000	8.16000	57.10000	36.00000	28.00000
1258	325 1500	29.00000	24.00000	5.98000	8.18000	57.10000	36.00000	28.00000
1264	325 1600	29.00000	5.00000	5.99000	8.17000	57.10000	36.00000	28.00000
1265	325 1600	29.00000	15.00000	5.98000	8.19000	57.20000	36.00000	27.90000
1266	325 1600	29.00000	24.00000	5.98000	8.17000	57.20000	36.00000	28.00000
1272	325 1700	32.00000	5.00000	6.10000	8.21000	57.20000	36.00000	27.90000
1273	325 1700	32.00000	15.00000	5.98000	8.20000	57.20000	36.00000	28.00000
1274	325 1700	32.00000	24.00000	5.95000	8.20000	57.10000	36.00000	28.00000
1280	325 1800	29.00000	5.00000	5.98000	8.22000	57.00000	36.00000	26.90000
1281	325 1800	29.00000	15.00000	5.98000	8.21000	57.10000	36.00000	26.90000
1282	325 1800	29.00000	24.00000	6.10000	8.21000	57.10000	36.00000	26.90000
1284	325 1900	29.00000	5.00000	5.95000	8.22000	57.10000	36.00000	28.00000
1285	325 1900	29.00000	16.00000	5.90000	8.21000	57.10000	36.00000	28.00000
1286	325 1900	29.00000	27.00000	5.85000	8.23000	57.00000	36.00000	28.70000
1288	325 2000	29.00000	5.00000	5.85000	8.23000	57.10000	36.00000	27.90000
1289	325 2000	29.00000	16.00000	8.88000	8.21000	57.10000	36.00000	27.90000
1290	325 2000	29.00000	27.00000	5.89000	8.29000	57.10000	36.00000	28.00000
1292	325 2100	29.00000	5.00000	5.85000	8.22000	57.00000	36.00000	27.90000
1293	325 2100	29.00000	16.00000	5.85000	8.25000	57.00000	36.00000	27.80000
1294	325 2100	29.00000	27.00000	5.90000	8.22000	57.00000	36.00000	27.80000
1296	325 2200	29.00000	5.00000	5.95000	8.22000	57.00000	36.00000	27.90000
1297	325 2200	29.00000	16.00000	5.95000	8.22000	57.10000	36.00000	27.90000
1298	325 2200	29.00000	27.00000	5.95000	8.22000	57.10000	36.10000	27.90000
1300	325 2300	29.00000	5.00000	5.90000	8.22000	57.00000	36.10000	27.90000
1301	325 2300	29.00000	16.00000	5.95000	8.22000	57.00000	36.10000	27.90000
1302	325 2300	29.00000	27.00000	5.85000	8.22000	57.00000	36.10000	27.90000

Figure 10

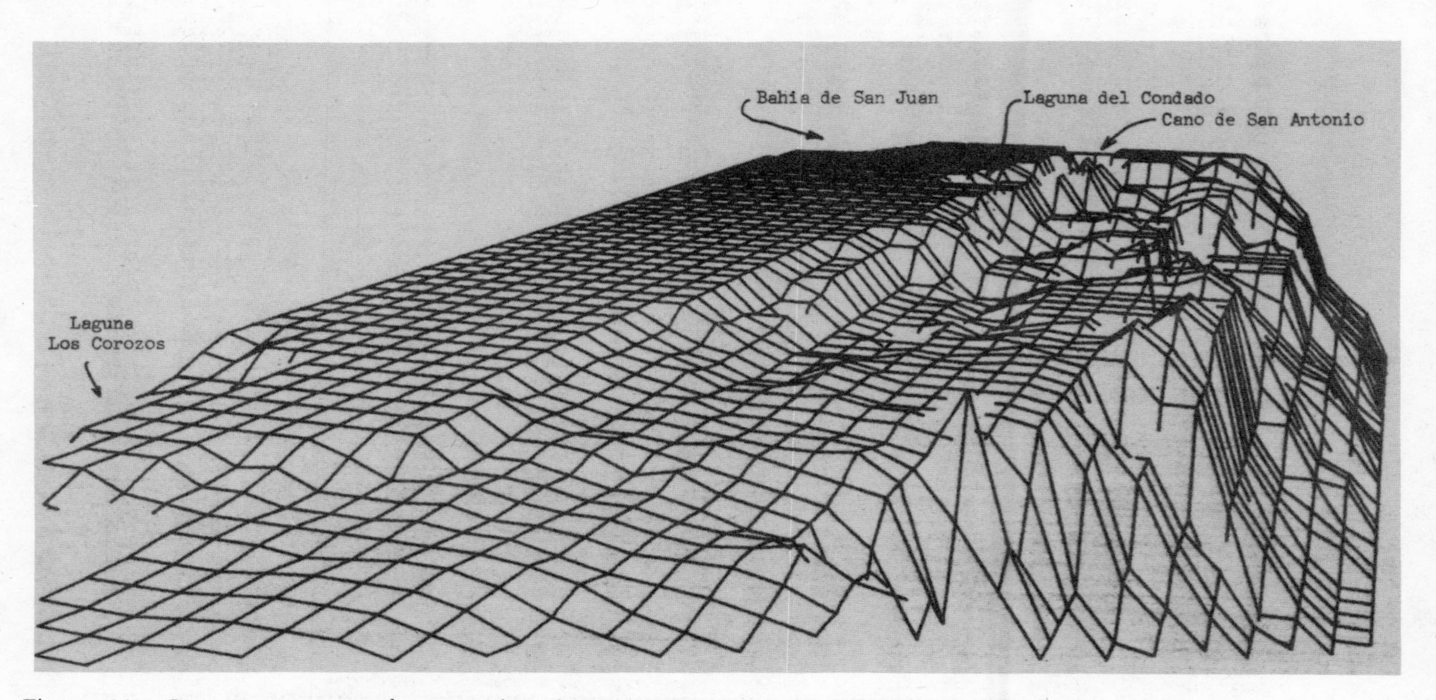

Figure 11. Computer generated perspective plot. Bottom topography off San Juan, Puerto Rico.

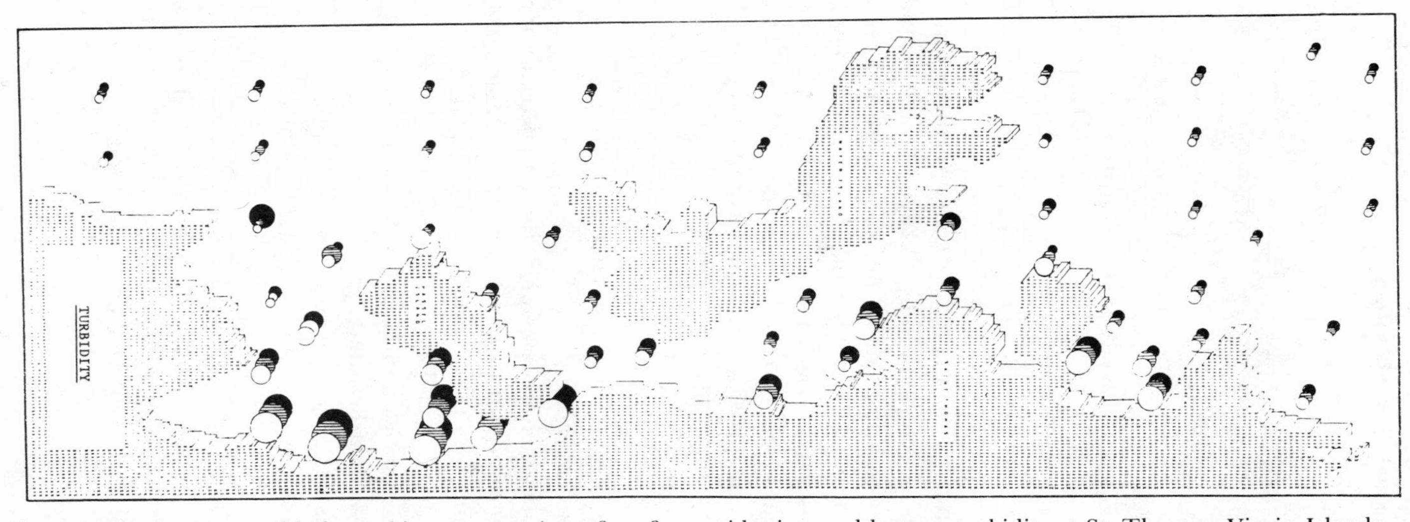

Figure 12. Computer aided graphics presentation of surface, midpoint, and bottom turbidity at St. Thomas, Virgin Islands.

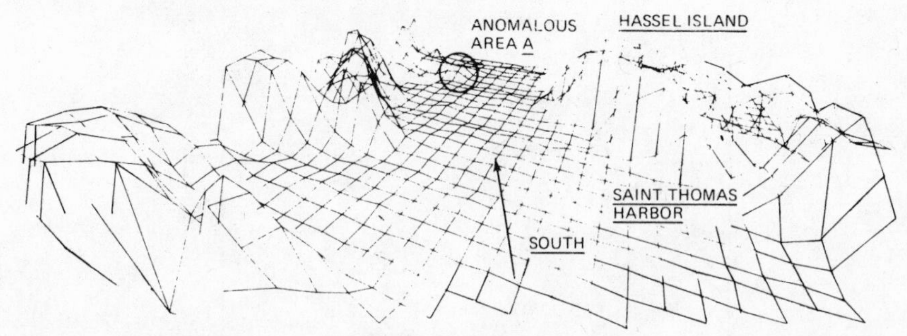

Figure 13. ERTS data, converted to perspective plot by digital computer.

quired, it is generally mandatory to compare them with aircraft and satellite findings to determine the effectiveness and boundaries of application of remote sensor data. When ground and sensor data are collected for a number of sample points, *correlation computer programs* can establish correlation coefficients to indicate the measure of association between variables. When a large number of variables exist, *factor analysis software* can provide a method of analyzing the intercorrelations in a set of variables. When there is high correlation between satellite (or aircraft data) and one or more ground factors intercorrelate with others, then remote sensor data will not only indicate specifics but may be used to provide clues to the status of other parameters in the total ecosystem.

Analog data-handling systems and the digital computer are invaluable in both transducer data processing and remote sensor data survey and mapping. Too often the role of data processing is underestimated: data acquisition and final data analysis are steps 1 and 3; they cannot, however, be totally effective without an adequate step 2, that is, environmental data handling.

This book was designed to give readers an opportunity to obtain an overview of modern environmental data-handling techniques. Master these techniques, for they serve as the fundamentals required by the successful data acquisition/data processing/data analysis and application team. Only the knowledgeable team can design a program that is both cost effective and of meaningful value to those decision makers who must act on the results of environmental data handling.

REFERENCES

1. *Symposium on Significant Results Obtained from the Earth Resources Technology Satellite—1.* Volume 1. NASA Publication SP-327.
2. *Third Earth Resources Technology Satellite—1 Symposium.* Volume 1. NASA Publication SP-351.
3. *Earth Resources Survey Systems.* Volume 1. NASA Publication SP-283.
4. *Symposium on Management and Utilization of Remote Sensing Data.* Proceedings, American Society of Photogrammetry.
5. *NASA ERTS-1 Experiment #589—St. Thomas, Virgin Islands, Final Report of October 1973.* W. C. Coulbourn, W. G. Egan, D. A. Olsen, and G. B. Heaslip.
6. *Earth Resources Program Results and Projected ERTS-1 Applications Investigations.* November 1973 report by NASA Lyndon B. Johnson Space Center.
7. *Machine Processing of Remotely Sensed Data—Short Course.* Laboratory for Applications of Remote Sensing, Purdue University.
8. *Remote Sensing with Special Reference to Agriculture and Forestry.* National Academy of Sciences, Book Number 309-01723-8, 1970.
9. *Earth Resources Technology Satellite Data Users Handbook.* Distributed by NASA Goddard Space Flight Center. General Electric Document No. 71SD4249.
10. *Remote Sensing of Earth Resources.* Volume 1. F. Shahrakhi, Ed. University of Tennessee, 1972.
11. *Interpretation of Aerial Photographs.* T. Eugene Avery. Burgess Publishing Co., Minneapolis, Minnesota.
12. *Applied General Statistics.* F. E. Croxton, D. J. Cowden, and S. Klein. Prentice Hall, Englewood Cliffs, New Jersey, 1967.
13. *Aerospace Telemetry.* Volume 1. Harry L. Stiltz, Ed. Prentice Hall, Englewood Cliffs, New Jersey.

14. *A Glossary of Range Terminology.* Range Commanders Council, White Sands Missile Range, New Mexico. Document No. AD 467 424.
15. *Measurement and Analysis of Random Data.* Julius S. Bendat and Allan G. Piersol. Wiley, New York.
16. *Proceedings of the American Society of Photogrammetry,* 41'st Annual Meeting. March, 1975.

GLOSSARY OF ENVIRONMENTAL AND TEST DATA HANDLING TERMINOLOGY

Alphanumeric. Refers to numerals, letters of the alphabet, or special symbols.

Analog-to-digital conversion (also A-to-D). The process of converting one or more analog samples (e.g., voltage levels) into a digital code.

Analog signal. One that varies continuously with respect to the variable it represents. Also the simulation of a continuous physical event by an infinite number of samples. (Digital simulations are based on a finite number of samples.)

Bandwidth. The difference between the two limiting frequencies (or wavelengths) of a band.

Binary. Two possible states. In digital recording a "1" or a "0" (pertains to the number representation system with a radix of 2).

Bit. Abbreviation for binary digit. Exists in one of two states: usually a "1" or a "0."

Calibration. Process of comparing an instrument with a "standard" to determine its accuracy.

Carrier. An electromagnetic wave that may be modulated to contain data.

Color composite image. Also referred to as color additive, color infrared, or false color. Usually a color print, negative, or transparency generated from three images from three different spectral bands.

COM. Computer output microfilm. Contains images of computer-output data routed to a cathode ray tube (CRT).

Commutation. Sequential sampling of several data parameters on a time-repetitive basis.

Compatible. Capable of accepting and processing digital data; for example, satellite computer compatible tapes (CCTs) are specially tailored for acceptance by most computer systems.

Computer (analog). A calculating device that solves problems in a continuous manner by representing physical variables by electrical quantities.

Computer (digital). A calculating device that processes information by combination of discrete and discontinuous data.

Contrast. A ratio of two adjacent scene radiances. Usually expressed as a number greater than or equal to 1.

Correlation. An analog or digital means of detection in which a signal is compared (point-to-point) with an internally stored known reference signal. Correlation is the degree of similarity between two waveforms or groups of data.

CRT. An abbreviation for cathode ray tube. When data (i.e., an electron beam) strikes the phosphor on the CRT face, that portion of the screen is energized and illuminated.

Demodulation. A process of retrieving an original signal from a modulated carrier.

Densitometer. A device that measures the optical density of each area of an image or photograph.

Digital. Expresses a value (of a data sample) in terms of a discrete number. Binary digital tape recording entails the transfer of binary data ("1's" and "0's") onto a magnetic data tape. Also pertains to the use of discrete integral numbers to represent a data sample's value.

Digitizer. A semiautomatic or automatic device that performs analog-to-digital conversion.

Dump (computer). A copy (e.g., printout) of all data in a storage device or on all or part of a magnetic data tape.

Electromagnetic radiation. In remote sensing commonly called radiation. Actually, energy propagated by sensors systems through space and detectable in wavelength bands.

ERTS. The U. S. Earth Resources Technology Satellite. First of a series, ERTS-1 was launched in July 1972.

File. Used by a computer to store types or groups of data.

Filter. A means of separating types of data or electronic signals based on frequency, amplitude, or other specified criteria.

FM discriminator. An electronic device that converts the variations in a carrier signal's frequency to analogous voltage variations.

Frame. Generally pertains to one complete commutator revolution.

Frame sync. A unique pattern that occurs at the start of each commutator period.

Frequency response. The portion of the frequency spectrum that can be sensed or recorded by a device without significant error.

Geometric accuracy. Pertains to *geographic accuracy* of data with respect to the earth's latitude-longitude system and *positional accuracy*; that is, the ability to locate a feature in an image with respect to a standard map.

Ground truth. Data pertaining to the physical state of earth surface features at the time of a satellite or aircraft overpass.

Hard copy. Printed copy of analog or digital machine output, usually on paper.

Hardware. Physical data processing equipment.

Hybrid. Digital and analog devices used together in a single system.

Imagery. A hard-copy record of the energy recorded by remote sensors. Usually a negative, positive, or photographic print.

Infrared energy. Heat radiation. Emitted by any object above temperature −273°C. All earth objects emit infrared radiation. Visible radiation appears at the lower limit of the IR wavelength interval; the upper end contains microwave radiation.

Instrumentation. Pertains to the science and art of design and use of instruments to measure and record the results of physical phenomena.

Keypunch. A keyboard-actuated device that punches holes in 80 column cards, each of which represents information; suitable for reading into a computer.

Landsat. The new name given to ERTS-2 following its launch in January, 1975.

Magnetic tape. An oxide-coated plastic tape (generally 7 or 14 tracks if the "instrumentation" type and 7 or 9 tracks if the computer type) on which signals can be recorded, stored, and retrieved.

Map (topographic). Shows correct horizontal and vertical positions of the features represented; uses contour lines to represent hills, slopes, and valleys.

Micron (μ). One millionth of a meter; a measure for wavelength.

Modulation. A means of placing information on a carrier such as AM or FM.

Multiband. Two or more sensors used simultaneously; each sensor measures radiation in a different wavelength band in the reflectance portion of the spectrum.

Multiplexing. Placement of two or more signals on a single channel (e.g., a tape track).

Multispectral. Two or more sensors used simultaneously to record energy in various wavelength regions of the electromagnetic spectrum. (Multispectral scanners are capable of recording IR data; multiband cameras are limited to the visible region.)

Noise. Undesirable electronic signals that mix with and degrade data. (Electrical storms, for example, degrade AM radio signals.) Also random variations of voltage, current, or data.

Ones complement. The radix (−one) complement of a numeral that possesses a radix of 2.

Oscillograph. A recording galvanometer system that converts electrical signals into a record on film or sensitized paper. Galvanometers contain a mirror that deflects a light beam onto photosensitive paper as voltage variations occur. A fiber optics oscillograph transfers a CRT image to paper or film.

Oscilloscope. A device with a cathode ray tube (CRT) that visually displays variations in electrical signals.

Parallel data. Digital words (i.e., containing 2 to n bits); each word bit is recorded simultaneously in tape recorder tracks $2 - n$.

Pattern recognition. A means by which shapes, forms, or specific features can be recognized by automatic means.

PCM. Modulation of a carrier by coded information pertaining to each data sample. Each PCM sample is coded with a combination of bits. Also pulse code modulation.

Program (computer). A sequence of coded instructions which act on data.

Radiometric. The nature of electronic and optical transmission of image data.

Radix. The integer used in a positional number system; all system numbers are expressed as sums of powers of that integer.

Real time. Pertains to transmission of data at the time acquired.

Registration (scene). In remote sensing the ability to superimpose a similar point in two different spectral band images taken at the same instant of time.

Remote sensing. The science and art of detecting and measuring characteristics of physical objects from a distance.

Resolution. The ability of a system to render visible the fine detail of an object or scene. In remote sensing resolution is generally specified in meters. A function of the signal-to-noise ratio and gain of a system. Ground resolution reveals the minimum distance at which a feature is detectable.

Resonant frequency. The specific frequency at which a given system or physical object will respond at maximum amplitude when subjected to an external sinusoidal force at constant amplitude.

Serial data. A train of bits, one following each other, for example, on a tape track.

Signal-to-noise ratio. Ratio of normal signal level to system noise level; usually expressed in decibels (dB).

Signature. The spectral response from an object or feature.

Software. The group of computer programs and routines associated with each computing system.

Spatial. Pertains to the location of an object based on linear dimensions.

Subcarrier. A carrier that modulates another carrier.

Subcommutation. Commutation of two or more signals to a single channel on a main commutator. (Subcom signals do not appear so often as do main commutator signals.)

Supercommutation. Connection of a single signal to more than one channel on a main commutator. When connected to *n* channels, *n* signals appear for each 1 on the main commutator.

Telemetry. System and means of measuring a physical phenomenon, transmitting measurement data, and recording that data at a distant receiving site.

Temporal. Relates to the study of physical data over time.

Terminal (computer). Hardware connected to a communication channel (e.g., a telephone line) which may input data or instructions to a computer and which receives the results of computer processing.

Track. Extends over the entire length of a magnetic tape but has a fixed position (and width) across the tape.

Tracking filter. An electronic device that attenuates unwanted signals and passes desired signals on an assigned frequency passband.

Transducer. A device that converts energy from one form to another (e.g., physical to electrical), such as the microphone.

Voltage controlled oscillator (VCO). An electronic device that converts voltage variations into a carrier which varies in frequency with respect to voltage changes.

Word (digital). A set of characters treated as one unit (e.g., a six-bit binary word can contain six bits such as 011011 or 000101). Each word or set of bits represents a data sample, synchronization sample, or calibration information.

INDEX